Peer-Led Team Learning: Evaluation, Dissemination, and Institutionalization of a College Level Initiative

INNOVATIONS IN SCIENCE EDUCATION AND TECHNOLOGY

Volume 16

Series Editor

Cohen, Karen C. *Weston, MA, USA*

About this Series

As technology rapidly matures and impacts on our ability to understand science as well as on the process of science education, this series focuses on in-depth treatment of topics related to our common goal: global improvement in science education. Each research-based book is written by and for researchers, faculty, teachers, students, and educational technologists. Diverse in content and scope, they reflect the increasingly interdisciplinary and multidisciplinary approaches required to effect change and improvement in teaching, policy, and practice and provide an understanding of the use and role of the technologies in bringing benefit globally to all.

For other titles published in this series, go to
www.springer.com/series/6150

Leo Gafney • Pratibha Varma-Nelson

Peer-Led Team Learning: Evaluation, Dissemination, and Institutionalization of a College Level Initiative

Springer

Dr. Leo Gafney
Independent Evaluator
Lakeville, CT
USA
leo.gafney@ sbcglobal.net

Prof. Pratibha Varma-Nelson*
Northeastern Illinois University
Chicago, IL
USA

*As of August 15, 2008
Professor of Chemistry and Executive Director of the Center for Teaching and Learning at Indiana University-Purdue University Indianapolis

ISBN 978-1-4020-6185-1 e-ISBN 978-1-4020-6186-8

Library of Congress Control Number: 2008930791

Printed on acid-free paper

9 8 7 6 5 4 3 2 1

springer.com

Acknowledgements

Not surprisingly Peer-Led Team Learning (PLTL) itself represents the learning we have gained from a large team of people committed to the improvement of undergraduate education in the sciences. We are grateful to our many colleagues from across the United States for their contributions to the evolution and growth of the project and to this study. In particular, we wish to thank David Gosser of the City College of New York, Director of the PLTL National Project, and Jack Kampmeier, senior colleague in the project, of the University of Rochester whose careful reading and many suggestions helped to improve the text. We acknowledge in a special way the peer leaders and students who have made invaluable contributions at every stage of development of PLTL and made it a pleasure to work on this project. Finally, we are most grateful to our spouses, Judy Gafney and John Nelson, for their encouragement and untiring support.

Introduction

There seems to be no end to the flood of conferences, workshops, panel discussions, reports and research studies calling for change in the introductory science courses in our colleges and universities. But, there comes a time to move from criticism to action.

In 1993, the Division of Undergraduate Education of the National Science Foundation called for proposals for systemic initiatives to change the way introductory chemistry is taught. One of the five awards was to design, develop and implement the peer-led Workshop, a new structure to help students learn science. This book is a study of 15 years of work by the Peer-Led Team Learning (PLTL) project, a national consortium of faculty, learning specialists and students. The authors have been in the thick of the action as project evaluator (Gafney) and co-principle investigator (Varma-Nelson).

Readers of this book will find a story of successful change in educational practice, a story that continues today as new institutions, faculty, and disciplines adopt the PLTL model. They will learn the model in theory and in practice and the supporting data that encourage others to adopt and adapt PLTL to new situations. Although the project has long since lost count of the number of implementations of the model, conservative estimates are that more than 100 community and four year colleges and a range of universities have adopted the PLTL model to advance student learning for more than 20,000 students in a variety of STEM disciplines.

This book is more than just a record of the PLTL story. Throughout, the authors distill out lessons of broader significance. For example, the six critical components for successful implementation of PLTL are pertinent to all efforts to effect educational change. The authors' analyses extend beyond local implementation to offer tactics for national dissemination and to suggest critical components of successful institutionalization of new pedagogies.

PLTL is a part of a significant shift in educational practices to provide new opportunities for student-centered active learning and the authors carefully situate PLTL in that larger context of change. On the other hand, PLTL is distinguished from many other initiatives by the central role of the peer leader. PLTL defines a new partnership with the faculty and staff and a leadership role for undergraduates that is appropriate to their abilities, while providing unprecedented opportunities to develop new levels of understanding of the discipline, and important teamwork, leadership, communication and interpersonal skills. When these gains for the leaders are added to those experienced by the students in the peer-led Workshops, the sum is a two-for-one result and compelling reason to pay attention to this insightful book.

Jack A. Kampmeier
University of Rochester

Contents

Chapter 1
Plan and Context of the Study

1.1 Background

Peer-Led Team Learning (PLTL) is an instructional model (http//www.pltl.org) that advances student achievement through active learning in a peer-led workshop (Varma-Nelson, 2006). During the past 15 years the method has demonstrated its effectiveness, improving students' academic performance in more than 20 studies. Conservative estimates are that PLTL is now used in more than 100 institutions—four-year colleges, community colleges, and research universities. More than 20,000 students, 150 professors, and 1,500 peer leaders are engaged in PLTL workshops each year, as an important part of their science courses. This study has grown out of more than 10 years of careful evaluation and monitoring of PLTL, and will provide detailed information, data, and references for all aspects of the project.The PLTL newsletter, *Progressions*, contains a wealth of information about the development and implementation of the project, and is available on the project website.

In a typical workshop, six to eight students meet with a peer leader for one and a half to two hours per week to discuss topics and solve problems that reinforce lecture and textbook learning, while also deepening their conceptual understanding and critical thinking. The workshop problems and activities are constructed to reinforce these goals and provide relevant applications. Most importantly, the workshops stretch students to work beyond what they could accomplish individually, so that through cooperative activities and appropriate guidance from the leader they reach new levels of understanding and performance. As students become more involved and engaged, they accept more responsibility for their learning and as a result their academic performance improves. The *PLTL Guidebook* (Gosser et al., 2001) explains the program, its theoretical foundation and offers practical advice for implementation.

The workshops are integrated into the course so that students can discuss their understanding of the concepts presented in the lectures and textbook in a non-threatening environment. Peer leaders facilitate the workshops, clarify goals, ensure that the team members engage with the materials and with each other, and they provide guidance as needed in solving problems. The process encourages collaboration and builds confidence. The leaders are students who have previously done well in the course and exhibit good communication skills and leadership potential. The workshop leaders

L. Gafney, P. Varma-Nelson, *Peer-Led Team Learning: Evaluation, Dissemination, and Institutionalization of a College Level Initiative*,

play an essential role because they are recent learners of the material. They relate to the students in the group as peers, understand how they learn, and explain material in ways that connect with them (Gosser & Roth, 1998). They are generally not responsible for grading student work, because it is important that they act as role models rather than authority figures. Unlike graduate teaching assistants, they have generally studied with the same instructor and from the same textbook as the students in their workshops.

Taken as a whole, the PLTL method forms a new pedagogy and consequently required careful monitoring and evaluation. The evaluation of PLTL was collaborative. Gafney, the external evaluator and co-author of this study, developed a theoretical framework for the overall evaluation, prepared instruments, studied implementation, identified outcomes, analyzed data, and prepared reports on various aspects of the project (Gafney, 2001a). The faculty members cooperated with the activities initiated by Gafney, collated and analyzed survey data and also conducted grade-based and standardized test-based studies of student performance (Tien et al., 2002; Wamser, 2006). Results of these studies will be presented in Chapter 2.

The first phase of the evaluation incorporated an approach similar to that described by Chen (2005) who identifies three stages of evaluation: analysis of implementation, monitoring by practitioners, and study of outcomes. In the evaluation of PLTL, implementation was analyzed using focus groups, questionnaires, and interviews to gather data about the project as the founding group of faculty began to use the method. Then the same faculty implementers and learning specialists (generally directors of learning assistance centers) monitored the program's progress at their institutions. They were primarily interested in student achievement of traditional learning goals measured by grades, and by the impact of the program on the peer leaders. This program monitoring also fostered increased faculty ownership, an interest in pedagogy, and a deeper understanding of PLTL-related issues, including an appreciation of the benefits of involving student leaders as partners in the educational enterprise. In the next phase, Gafney analyzed PLTL outcomes across sites, comparing implementation strategies, identifying problems, and analyzing how adoption of PLTL workshops interacted with departmental and institutional cultures.

In addition to being collaborative, the evaluation was multidimensional, looking beyond student performance, to consider each of the following: quality of initial implementation, the effect on leaders, and adaptation issues by type of institution and discipline. The evaluation also studied dissemination of the method particularly through mini-grants. Finally the requirements for institutionalization were studied.

1.2 Plan of the Study and Methodology

This was a broad-based study. The goal was to look closely at implementation, dissemination, and institutionalization in order to identify key factors that would be transferable to other educational innovations. The study considers three evaluation questions: (1) What is required for a new approach to teaching and learning to be successfully implemented at the college level? (2) What is required for dissemination of the method across disciplines and institutions? (3) Finally, what are the critical elements needed to successfully institutionalize a program?

This study therefore has several parts. First, how is a model for teaching and learning developed and tested? Peer-led workshops are embedded in theories of learning and ideas about instructional practice, but they are also rooted in the cultures of departments and institutions. Workshop programs were developed, piloted, assessed, and revised. But this did not happen in a vacuum. Institutional and professional priorities had to be addressed and this led to modifications. We found that the type of institution, the discipline, and previous history were significant factors in the implementation and in the success or failure of PLTL at each site.

The second part of the study looks at dissemination issues from the perspectives of the disseminators and of the recipients. The PLTL project developed a four-tier model that proceeded from creating initial interest in the method, to assisting with implementation, and exploiting dissemination opportunities. Peer-led workshops were adapted to local circumstances and needs, and frequently were introduced with other new approaches to teaching and learning. These adaptations introduced other variables but they also made the results more interesting and valuable.

The third part of the study identifies critical success factors required for the institutionalization of PLTL. We began with a set of hypotheses that included: adherence to the model, fit with local needs, and administrative support. We found that these were important but data analysis uncovered other important sustainability factors related to faculty cooperation, motivation, and adaptation of the model.

The three parts of the study just described are covered in Chapters 2 on implementation, 3 and 4 on dissemination, and 5 on institutionalization. Chapter 6 presents the results of a careful study of the impact of PLTL on peer leaders, as they looked back on the experience from the vantage point of up to 10 years. Chapters 7, 8, and 9 treat particular areas related to the program—under-represented minority students and women, new paradigms for teaching and learning, and special issues.

Results of the evaluations are presented in various parts of this study, as appropriate. In Chapter 10 they are collected and reviewed with suggestions for adapting them to other projects. It is important to note that the strategies used and data collected grew organically out of the project, as it evolved over the years and grew in complexity. Methods used in this study included the following.

- *Surveys*. Since the first years of PLTL, surveys have been employed with students, faculty, and peer leaders to gather information about their experiences and their satisfaction with the program. An online survey with former leaders was used to gather data about the impact of leading workshops on the leaders' further studies and first career steps. Another online survey of PLTL faculty was used to gather data about the perceived success of PLTL, dissemination activities, and institutionalization.
- *Interviews*. Over the years of the PLTL grants, interviews have been conducted with the students, peer leaders, faculty, and administrators to gather information about experiences with the program, problems in implementation and dissemination,

the effectiveness of workshops, benefits to students, and other important issues such as funding. These interviews have generally been semi-structured, with the same questions being asked of a population of faculty, students, or peer leaders, but with room in the interview to pursue areas of individual interest. Interviews have generally lasted about 30–40 minutes, were recorded, transcribed, coded according to need, and analyzed.

- *Focus groups and discussions with workshop students and leaders*. Discussions with those engaged in workshop learning provided a rich source of insights into the perceptions of students about their own learning as well as the academic, social, and motivational impact of workshops on students.
- *Comparative studies and other statistical measures*. Instructors using PLTL have collected grade data based on students in classes with and without workshops. They have also collected scores from standardized tests for cohorts with and without workshops. Some instructors have determined the impact of workshop attendance, comparing numbers of workshops attended with grades.
- *Review of the literature on learning and academic support programs*. Both Gafney and faculty members associated with the project have studied the literature surrounding peer-led workshops. Insights into learning theory, pedagogy for small groups, developmental stages, student diversity, other reform initiatives, and dissemination have been important in assessing the progress and potential of PLTL.
- *Site visits*. During the years of the project and as part of the supplemental grant on dissemination and institutionalization, the project evaluator, Gafney, made numerous visits to PLTL sites, interviewing faculty and leaders, observing workshops and talking with students to identify the keys to successful implementation, dissemination, and institutionalization.
- *Participant observations*. One of the authors, Varma-Nelson, has given numerous workshops for faculty members, as well as formal presentations at conferences and seminars throughout the United States and in several foreign countries. These experiences have yielded insights into the issues and problems associated with adopting, adapting, and implementing workshops.
- *Administration and data collecting related to mini-grants*. Varma-Nelson was the administrator of 92 mini-grants to PLTL adopters. The applications for these grants, reports submitted, and responses to phone interviews and site visits yielded a wealth of data about all three areas of the study: implementation, dissemination, and sustainability. We also gained a deeper understanding about the use of mini-grants as a dissemination strategy.

1.3 Origins and Rationale for Peer-Led Team Learning

The organizational arrangements surrounding college-level science courses are well established. Lecture hours, textbooks, quizzes and exams, supplemented in some cases by labs and recitations, define most courses. The assessment methods define what students focus on. There is, of course, variety in lectures and in textbooks. Some professors are clearer, more interesting, and bring the material to life more than others. Students, however, often find themselves more passive than active in lectures.

A number of teaching/learning problems encouraged the founding group of instructors to try Peer-Led Team Learning. Among these were:

1. Professors of chemistry at the participating institutions described how general and organic chemistry had changed in recent decades. Students are now required to engage in material that is more conceptual, quantitative, and challenging than in the past. Whereas memory once sufficed in passing science courses, these professors pointed out that rote memorization no longer works. Students have to understand concepts in order to solve problems.
2. Professors at many institutions find that even successful students frequently do not know how to communicate scientific ideas or work on problem solving teams.
3. Many students do not seek the help they need nor do they utilize the on-site resources available to them.
4. A large number of students are not actively engaged in their own learning. Faculty members recognized and accepted the fact that the handing down of knowledge must be complemented by individual and social learning activities that promote intellectual and personal growth. They also saw that many students were simply not “getting it,” in lectures. They were therefore looking for approaches that would supplement lectures, requiring students to be more intellectually active.

In addition to these personal experiences and reflections, the project is supported by the literature on why students are challenged and why many leave the sciences. Studies point to a range of reasons for attrition. First, according to some, there is a tendency on the part of faculty—implicit or explicit—to blame students for their failures (Lovitts, 2001). These studies indicate that a variety of factors beyond student effort and achievement are involved and should be considered. Tobias (1990) described differences in the way minority students tend to approach learning and the fact that many pedagogical innovations are of particular benefit to these students. Seymour and Hewitt (1997) provided evidence that many students leave the sciences because of problems, “which arise from the structure of the educational experience and the culture of the discipline” (p. 392). These problems are related to, “pedagogy, student assessment, curriculum design and advising” (p. 394).

Astin and Astin (1993), in a major longitudinal study of more than 27,000 students, found that adequate preparation in mathematics was the single best predictor of students persevering as science majors and entering science-related careers. In a related study, Astin (1993) concluded that “the student’s peer group is the single most potent source of influence on growth and development during the undergraduate years” (p. 398).

The PLTL founding group saw that many of the concerns outlined above could be addressed by supplementing their lectures with PLTL workshops. Students would spend more time problem solving, become more active in their learning, communicate more effectively with one another, review the lecture material, have an added opportunity to ask questions and test their understanding, and think more deeply about the conceptual side of their learning (Tien et al., 2002; Varma-Nelson, 2006). In addition, Varma-Nelson and Coppola (2004) cite group learning, reciprocal teaching, studio instruction, and social constructivism as important areas of educational design contributing to the strength of PLTL.

1.3.1 PLTL Within the Institution

As was mentioned above, colleges tend to preserve their organizational structures. Requirements for graduation, for a major, and for courses are well defined and don't change much from year to year. So the introduction of a new pedagogy or requirement can encounter resistance within the department, and this has sometimes been the case with peer-led workshops.

A college professor has autonomy within his or her course with regard to lecturing technique, selection of texts, construction of exams, and grading. Introducing peer-led workshops, however, took several steps beyond the usual course expectations. Additional time and space were needed for workshops; involved faculty had to develop materials; leaders needed to be trained; and funds were required to pay the leaders. These were not monumental issues but each area required negotiations within and often beyond the department. The capacity of PLTL to overcome barriers and gain traction varied with the local implementer, the institutional climate, departmental access to funds, and other variables. After initial implementation, students and peer leaders became important factors, helping the project gain greater visibility. At some sites, students experienced the workshop benefits immediately and their voices were heard by administrators and faculty. At other institutions even with evidence that the workshops were improving student performance, administrators were not persuaded that the program was worth the effort and expense.

1.4 PLTL in the Context of Reform Initiatives

We describe several educational reform initiatives below to provide a context for the present study. We will analyze these in greater detail in Chapter 5 which is devoted to institutionalization.

In recent years we have witnessed an explosion of pedagogical initiatives that foster more active learning. There are data supporting the fact that students can reap significant academic gains from: peer tutoring (Miller et al., 2001), study groups (Light, 2001), and cooperative learning (Dubinsky et al., 1997; Johnson et al., 1991). Programs that stress active learning are springing up at colleges everywhere. Examples include: Merit workshops (Treisman, 1985), Supplemental Instruction, Excel Workshops, Peer Instruction (Mazur, 1997), Guided Inquiry (Moog, 2004), and many more. Vast amounts of information can be found about each of these and related programs on the web, at conferences, and in the literature. Higher education has been working to assimilate these findings.

PLTL has entered the mainstream of reform efforts so that it is now a recognized force within the world of college teaching and learning in the sciences. We will now look at several initiatives in greater detail to preview the issues and potential problems involved in implementing and sustaining a pedagogical initiative.

In 1968 Fred Keller introduced the Personalized System of Instruction (PSI) for college level instruction. As explained by Cracolice and Roth (1996) this approach allowed students to move toward mastery of the material at their own pace. The method employed proctors for scoring and tutoring. Lectures were reduced to a minimum or even eliminated. The method was widely adopted and found to be effective in chemistry courses, with a high level of student and instructor satisfaction. But over time, the number of new adoptions declined and many sites that had adopted the method, later abandoned it. The method is now little more than a pedagogical footnote. According to Cracolice, the major issue seems to have been the amount of time and effort required to maintain the individualized arrangements.

Reform Calculus was developed in the mid 1980s to address a number of learning issues including the mindless use of algorithms and students' inability to employ the methods of calculus in subsequent courses (Tucker & Leitze, 1995). Reform Calculus seemed to be the right thing at the right time. Instructors wanted a less formal and more intuitive approach and one in which calculus would not be such an ogre in the minds of students and such a gatekeeper in the minds of faculty. Aided by grants, new materials, and adoption at prominent universities, Reform Calculus took hold. Although there were adversaries, the newer approach is now used to a greater or lesser extent at colleges across the country. Success was probably due to the fact that most basic structures such as lecture time and testing remained unchanged, but the program did address a felt need to make learning more relevant.

The explosion in the use of technology can also be considered as an educational initiative. In mathematics, the graphing calculator and computer programs can be important aids to understanding by eliminating tedious calculations, revealing the effect of changes in parameters, and demonstrating the behavior of functions—to mention a few benefits. The computer has also become a major force in science and math courses—as an aid to course management and presentation, and as a tool for analyzing data. New technology has been successful because its use opened new doors to learning.

The expansion of undergraduate research in the sciences is an educational initiative that has gained momentum in recent years. Funding is available, lab directors are happy to have the help, and everyone sees important benefits as students begin to do science, not just learn about it (Seymour et al., 2004; Gafney, 2001b).

Finally, the emergence of inter-disciplinary courses, programs, and majors has created situations in which faculty members must cooperate and students are involved in new kinds of learning. Bioinformatics, Environmental Physics, and Urban Ecology are but a few of the many new interdisciplinary offerings causing faculties to rethink their approaches and providing new opportunities for students. Interdisciplinary courses provide solutions to problems and answers to questions that would not be possible within the constraints of a single discipline.

There are some general lessons to be gained from these reform efforts.

- *Problems and needs*. Faculty will engage in a reform activity or adopt a new approach if they see it as solving a real problem or answering a real need. Both reform calculus and computer-related technologies led to more active learning. Students became more interested because they could "see the point" of what they were doing. Interdisciplinary approaches are required when a problem's solution depends on tools from another area.
- *Costs*. Faculty must perceive the benefits of the program as worth the different costs. With regard to the Keller plan, it seems instructors found the investment required—in the development of materials, supervision of proctors, and individualized testing—to be more burdensome than the potential benefits, and so they abandoned the method. Undergraduate research is often funded by grants, provides an important service, and uses existing structures; and thus is not difficult to initiate. Reform Calculus and the use of technology required faculty time and energy but this investment could be tailored to the instructor's resources and commitment, and the expenditure diminishes over time.
- *Professional fit*. Faculty adopting the reform need to see the activities as professionally compatible with what they generally do and how they define their roles. Some instructors have been averse to Reform Calculus because they see it as a betrayal of mathematics which should, in their view, place a high priority on formal proof and rigorous procedures. Introduction of a new course is generally not difficult if there is a faculty member interested in teaching it and enough students wish to enroll. Initiating a new department or major often takes years of effort in which institutional requirements must be satisfied. Finally, the fit with a professor's other responsibilities, particularly the research commitment, is critical and will be discussed at several points in the study.

This study of Peer-Led Team Learning will look into these and other areas as we consider the development of the model, implementation of the method, dissemination, and institutionalization. In each of these areas a variety of incentives and disincentives presented themselves.

Tensions were found in each phase of the PLTL project. We found that in successful programs a balance was struck between competing demands. Professors wanted to engage and motivate students but they had a limited amount of time to invest. Peer leaders can add a dynamic dimension to the teaching/learning process but they require training and supervision. Adaptation helps new sites gain ownership but a program can lose its effectiveness when stretched too far. The PLTL project faced an ongoing challenge in addressing the needs of faculty and institution while maintaining the integrity of the program as defined by the critical components.

In summary, the goals of this study are to document and analyze the history of Peer-Led Team Learning over a 10 year period. We have a wealth of data covering each stage of the project, and will present these data along with appropriate theoretical information about teaching and learning, dissemination, and sustainability. In addition to successes and failures of PLTL we will also discuss the process of educational change at the college level, deriving lessons applicable to other initiatives.

Chapter 2
Development and Implementation of Peer-Led Team Learning

In this chapter we describe and explain how PLTL developed as a program. The process was one of organic growth and teamwork. The end result was not seen as a package or even a product. The program was, and is still, a work in progress.

There were several important events as the founding group implemented and analyzed the program. The first event was the development of six critical components required for successful implementation of PLTL. These elements provided a framework and vocabulary that gave the program stability, and facilitated dissemination. The second important factor was the commitment to measure the impact of the program on student performance. Almost every implementer generated sets of data comparing student outcomes with and without the PLTL. Third, there was among the founding group a recognition that the peer leaders were key to the success of the program, and so great care was taken to select, train, and monitor these leaders.

We will explain the development and implementation of PLTL with attention to these three aspects of the program: development of a model, generating data, and the special role of the workshop leaders.

2.1 Peer-Led Team Learning as a Program

A program can be defined as a set of resources and activities directed toward intended outcomes (Wholey, 1994). This definition is helpful in examining PLTL which fits nicely into the categories of: resources, activities, and outcomes.

The resources brought to bear in PLTL include: the experiences, skills, and goals of faculty members; institutional priorities; funding; educational materials and facilities; former students available as leaders and current students. In the area of faculty resources there are a number of variables: whether an individual or small group is considering implementation; the number of faculty members who are involved; institutional history of concern with pedagogical issues; and the level of commitment on the part of potential implementers. At the administrative level, familiarity with the program and commitment to the improvement of teaching and learning are key variables.

The activities needed for the successful launching of PLTL are not large in number but critically important in quality for the program to succeed. The most

L. Gafney, P. Varma-Nelson, *Peer-Led Team Learning: Evaluation, Dissemination, and Institutionalization of a College Level Initiative*,

important activity is of course the workshop itself and all other activities are in support of the workshops. These activities include: selection, training, and on-going supervision of leaders; preparation of materials; plans to pay student leaders; and arrangements regarding time and space.

When workshops were first implemented, there were a number of desired outcomes. These included: improved student performance and interest, increased retention in courses, more active learning, enhanced problem solving abilities; and the development of team skills.

2.1.1 Initial Implementation of Peer-Led Team Learning

The elements of a program, as outlined above, were implicit as the PLTL founding group began to organize student led workshops in the early 1990s. In 1995, Workshop Chemistry—the precursor of PLTL—was one of four recipients of the National Science Foundation grants for systemic change in chemistry. The Workshop Chemistry consortium was led by the City College of New York (CCNY) and included 10 colleges of the City University of New York (CUNY) system as well as St. Xavier University in Chicago, and the Universities of Pittsburgh, Pennsylvania and Rochester. Workshops were initiated at each of these sites and experiences were positive.

This grant was followed by NSF, CCLI (Adopt and Adapt), grants for the implementation of PLTL with two consortia. Participants in the first consortium were the University of Montana, American University, Clark Atlanta, and the University of Kentucky. The second consortium included: The University of West Georgia, Miami of Ohio, Coastal Carolina University, Indiana University Purdue University at Indianapolis, Goucher College in Baltimore, and Prince George's Community College in Largo, Maryland, outside Washington, DC.

An NSF grant was then awarded to disseminate the PLTL workshop method geographically throughout the nation, and across the various science disciplines—particularly in biology, physics, and mathematics. This grant was augmented by a supplement supporting activities to disseminate the model at community colleges. A second national dissemination grant was awarded, followed by a supplement used to complete the work supporting this report.

A key strategy supported by both dissemination grants was the use of mini-grants, entitled Workshop Project Associates (WPA). Throughout the project, 92 such grants were awarded. Activities, outcomes, and issues associated with this mini-grant program are analyzed in detail in Chapter 4.

2.1.2 Development of a Model

One of the key events during the first years of the project was the identification of a model containing six critical components for success. The external evaluator, Gafney, proposed elements of the model, but all of the initial PLTL implementers accepted these and assisted in the development of the model. These components have been repeatedly found to contribute to successful student performance. They also provided a solid basis for evaluating implementation and monitoring progress. The critical components are:

1. *Faculty involvement.* The faculty member teaching the course is closely involved with the workshops and the workshop leaders. As distinct from Supplemental Instruction and tutorial programs run from an academic support center, the workshops are usually planned and implemented by the professor teaching the course. Most often, he or she is responsible for the selection and training of leaders and in the development of workshop materials.
2. *Integral to the course.* The workshops are an essential feature of the course, not an optional add-on. For the workshops to be effective it is important that they be seen by students as belonging to the course, as important as lecture, homework assignments, and tests. For most of the original instructors this meant that the workshops were mandatory. With the passing of time and diffusion of the program, for one reason or another, many instructors have made workshops a matter of student selection, but as a decision made at the start of the course—not as drop-in sessions. It is, however, essential that both students and instructor view the workshops as highly significant in student learning—a requirement at least for those who have selected to participate.
3. *Leader selection and training.* The workshop leaders are carefully selected, well trained and closely supervised, with attention to knowledge of the discipline and teaching/learning techniques for small groups. Very early in the development of PLTL it became evident that the peer leader's role was the distinguishing characteristic, and that the "peer" quality was as important as that of leader. Group problem-solving sessions are common among college students but the presence of a trained leader is not. In the earliest focus groups in 1994, students talked about positive feelings in workshops, not worrying about asking a "stupid question," and that the leaders could sometimes explain things in ways that complemented the lectures, while professors often answered questions by "explaining things the same way," a second time. Leader training from the start included attention to: content of the discipline, the learning process, and the individual participants.
4. *Appropriate materials.* The workshop materials are challenging, intended to encourage active learning and are suitable for group work. The development of appropriate materials is critical to the success of the workshops. Most textbook problems and exercises are not useful for the workshops, without revision, because they are written for individual rather than small group work. The ideal workshop problems need to be sequenced so that they move seamlessly from easier to more difficult material, reinforce the lectures, and work effectively with small groups. As the project proceeded, materials were developed and eventually published for general and organic chemistry (Gosser et al., 2006; Kampmeier et al., 2006; Varma-Nelson & Cracolice, 2001). These were often mentioned in site visits and reports from adopters as factors facilitating implementation.

5. *Appropriate organizational arrangements*. The particulars, including the size of the group, space, time, noise level, etc. are structured to promote learning. Implementers have repeatedly found that the ideal group size for PLTL workshops is six to eight students. With fewer students it is difficult to get enough diversity of ideas and responses or to generate cooperative work. With more than eight, the group tends to fragment or to take on the characteristics of a recitation, and it is difficult for the peer leader to provide direction. Similarly, 90–120 minutes has been found to be the best length for the workshops. Less time has not proven adequate for productive cooperative work and the development of problem-solving skills.
6. *Administrative support*. Workshops are supported by the department and the institution. This component was added to the list after a year or so of experience when it became apparent that the workshops could not achieve even a basic level of successful implementation, without support. This support includes a financial commitment by the institution and a level of positive reinforcement for those taking the time to implement the program.

2.2 Evaluation Priorities

Bodner (1999) distinguishes educational assessment from evaluation. He considers assessment to cover the approaches and outcomes used in judging the success of individual students in learning, usually conducted through the use of tests and grades. Evaluation is a larger activity that is used to make decisions about changes in teaching and learning. Assessment should be used as part of a total evaluation, but student grades or other academic outcomes are not in themselves a basis for curriculum or teaching and learning decisions. He compares the process to sports. If a baseball team wins a game by a score of 10 to 1, we know who won the game. But the score in itself does not provide enough information for a manager to decide who will play in the next game. Bodner goes on to discuss approaches to evaluation, pointing out that there are many important questions about educational initiatives that will not be answered simply by testing students. Some of his questions could have been written with PLTL in mind and were in fact very similar to the questions we were asking at various times in the evaluation. They include:

- How do we overcome student resistance to this approach?
- What factors interfere with the ease with which this technique can be used by other instructors, or transported to other institutions?
- What effect does this mode of instruction have on the instructor's attitude toward teaching?
- Does this approach to instruction produce students who are more likely to think the way a scientist thinks?

Coppola et al. (2001) discuss similar priorities in listing six forms of assessment used with structured study groups: exam performances; numerical and narrative survey information; leader reflections; counterintuitive follow-up assessments; student reflections; and a course portfolio. Like these evaluations, PLTL was interested in many aspects of teaching and learning. We were interested not only in outcomes but the conditions favoring or inhibiting desired outcomes. As a result we had to throw the evaluation net very wide in order not to miss any aspect of the program. Nevertheless, we shall see later in this chapter that many comparative studies have demonstrated positive effects of PLTL on student learning as measured by tests and grades

2.3 Students and Student Leader Experiences: Qualitative Data

During the 1996–97 fall term the first wave of PLTL students and leaders completed surveys about the workshops. Written questionnaires were sent to the participating institutions to obtain data about students' experiences in the workshop courses. Survey items asked about relationships with the workshop leaders and with other students, involvement of the faculty, and the materials and arrangements used for the workshops. Workshop leaders were asked similar questions and others about their training and support in conducting the workshops, and whether they viewed the workshops as generally successful.

Nine institutions and at least 16 different classes were represented in the survey. Responses were tabulated according to institutions. The size of the student sample reporting was generally greater than 50 percent and frequently nearly 100 percent of those in a workshop course.

Table 2.1 shows the numbers of responses from students and student leaders reporting, with 9 groups of students and student leaders.

2.3.1 Student Responses

Percentages of agree, neutral, and disagree responses were tabulated first for each group. Then overall averages were computed by calculating the average of the

Table 2.1 Number of survey responses by institution

Institution	Class	Students	Leaders
City College	General Chemistry	262	7
University of the Pacific	Organic	28	4
University of Pittsburgh	General Chemistry	115	14
St. Xavier	Organic	25	5
NYC Technical C.	General Chemistry	34	6
University of Rochester	Organic Chemistry	166	25
University of Pennsylvania	Organic Chemistry	30	3
Medgar Evers	General Chemistry	46	6
Borough of Manhattan CC	General Chemistry	17	5
Total		**723**	**75**

group means. Each group is thus considered as a separate sample. Nevertheless, the differences between group averages and total population averages were small.

In general, students found the workshops helpful to their learning. Items 3, 4, 9, 10, 15, and 16 are positive statements about their learning and were all agreed to by more than 75 percent of the respondents, while less than 10 percent disagreed with any of these items. These items contain statements that: interacting with the workshop leader helps learning; the materials are demanding, integrated with the course, and useful; that the respondent would recommend the workshop course to friends; that they are comfortable asking questions; and the leader is well prepared.

Responding to Item 6, 70 percent of the students agreed that the workshops increased their grade, while 13 percent disagreed. There was rather wide variance among group averages to some items. Average agreement with Item 1 stating that "the course as a whole is well organized," ranged from 50 to 91 percent. Average agreement with Item 2 stating that "the lecturer clearly presents the chemistry," ranged from 37 to 100 percent. Many students who reported that the course was not well organized or that the professors were not clear, nevertheless spoke very favorably about the positive impact of PLTL on their learning. Responses for items in the following table (Table 2.2) were: 5 = strongly agree; 4 = agree; 3 = neutral, no opinion; 2 = disagree; 1 = disagree strongly.

Table 2.2 Workshop Chemistry Student Questionnaire: fall 1996

	% Agree	% Disagree
1. The course as a whole is well organized	76	10
2. The lecturer clearly presents the chemistry	69	15
3. Interacting with the workshop leader increases my understanding of chemistry	81	6
4. The workshop materials are well connected to the lectures	85	7
5. My workshop group sometimes has extra meetings to prepare for tests or to review difficult material	31	51
6. I believe that the workshops are improving my grade	70	13
7. I regularly explain problems to other students in the workshops	50	18
8. Interacting with the other group members increases my understanding of chemistry	72	8
9. I would recommend workshop courses to other students	82	7
10. In the workshops I am comfortable asking questions about material I do not understand	86	5
11. The lecturer encourages us to participate in the workshops	80	7
12. Noise or other distractions made it difficult to benefit from the workshops	21	61
13. Students who are uninterested or unmotivated made it difficult for others to benefit from the workshops	21	59
14. The workshops are not helpful because I already know almost everything that is covered	6	84
15. The workshop materials are demanding and are good preparation for the tests	77	8
16. The workshop leader is well prepared	75	6

2.3.2 *Leader Responses*

The table below (Table 2.3) shows overall averages of leader responses. In this case the groups are aggregated and averaged over total number of leaders. Differences between the averages calculated this way and calculated by groups did not differ by more than a few percentage points except in the case of Items 16 and 17 in which the averages by group are about 10 percentage points higher than those by individual.

As in the case of students, the leaders' responses indicate highly positive experiences with the course and the workshops. Agreement with Item 8, "I would recommend workshop courses to other students," is even greater for leaders (100 percent) than in the case of students (Item 9, 82 percent).

The leaders also agreed, by a wider margin than students, that the workshops improve student grades (leaders, 91 percent; students, 70 percent). The leaders are much more involved and have a greater emotional investment than the students. Consequently, it might be expected that as a group they consider the workshops more effective than students do.

Regarding Item 3, "Acting as a workshop leader increases my understanding of chemistry," 97 percent agreed, while none disagreed.

Table 2.3 Workshop chemistry: Peer Leader Questionnaire: fall 1996

	% Agree	% Disagree
1. The course as a whole is well organized	89	0
2. The lecturer clearly presents the chemistry	74	0
3. Acting as a workshop leader increases my understanding of chemistry	97	0
4. The workshop materials are well connected to the lectures	90	0
5. My workshop group sometimes has extra meetings to prepare for tests or to review difficult material	41	30
6. I believe that the workshops improve student grades	91	0
7. I regularly explain problems to students in the workshops	91	2
8. I would recommend workshop courses to other students	100	0
9. In the workshops students are generally comfortable asking questions about material they do not understand	98	0
10. The lecturer shows an interest in me as a workshop leader	80	0
11. Noise or other distractions sometimes make it difficult to benefit from the workshops	39	53
12. Students who are uninterested or unmotivated make it difficult for others to benefit from the workshops	38	40
13. Interacting with the other workshop leaders is helpful	90	5
14. The workshop materials are demanding and are good preparation for the tests	90	0
15. Students are generally well prepared for the workshops	34	41
16. As a workshop leader I act more as a guide than a teacher	77	6
17. The training that I have or am receiving on how to conduct workshops is helpful	79	0

Responding to Item 16, "I act more as a guide than a teacher," 77 percent of the leaders agreed. Similarly, Item 17 stating that leader training was useful received 79 percent agreement.

About 38 percent of the leaders agreed that "noise and other distractions make it difficult to benefit from the workshop," and that "uninterested and unmotivated students make it difficult for others to benefit from the workshops," while only about 21 percent of the students agreed with these statements. It may be that the leaders were more distracted than the students by the external circumstances of the workshop and by unmotivated students.

After the initial surveys, tabulated above, the project evaluator made revisions, and the surveys were made available to sites implementing the method and many of these used the surveys as part of their internal evaluation, moving into a phase of program monitoring by participants.

In the years since the initial implementation of PLTL, some institutions receiving grants to improve teaching and learning have requested further analysis of student experiences, based on interviews and open-ended survey items. In 2003, PLTL was introduced at Brooklyn College, of the City of New York, in conjunction with an NIH-sponsored project for under-represented minority students. These students were selected for the program because they were judged to have potential but to need some extra support in college. Surveys, interviews and group discussions uncovered a number of areas in which students felt workshops were beneficial to them. These included the following.

- *Understanding*. Interacting with the peer leaders brought the subject matter into the students' world and experience. "They give you the impression that learning is not some crazy difficult thing but actually makes sense."
- *Confidence*. Many of these students were apprehensive starting college and success in the workshops helped them believe that they could do the required work. "Workshops created a sense of confidence and well-being, that assured me I can take on any challenge."
- *Group work*. We have often been told by students that when they encounter obstacles in their learning they tend to give in to distractions and leave the assignment. Working in a group keeps them on task. "Working with a group helped overcome difficulties I was having on my own."
- *Problem solving*. "Workshops provided opportunities to express myself, and examine myself to see if I was on the right track." As teachers at all levels know, a student can proceed a long way in the wrong direction without knowing it. The group helps channel problem-solving skills and energies.
- *Leader and individual work*."The workshop leaders have shown me that to help myself I must push myself. No one is going to do it for me. Workshops have helped me become more responsible in doing my work."

2.4 Academic Performance: Quantitative Data

Instructors implementing Peer-Led Team Learning were generally pleased to note that students were more motivated and more interested in the course as a result of the workshops, and the trends noted in the earliest questionnaires were generally reproduced by later adopters. But instructors also wanted evidence that student performance improved as a result of the program. Most felt that the best evidence would be grades and test data. Most had not changed the test material from the years immediately preceding the introduction of PLTL, and so they had an historic basis for comparing student performance. Based on college entrance data, these instructors were also confident that the cohorts in their courses had not undergone significant changes within the four or five years prior to the adoption of PLTL. Other professors started with only some students engaged in workshops, and so they had natural comparison groups based on the course structure.

The professors interested in evaluating PLTL did not have the luxury of what Chen (2005) calls efficacy evaluation. This method is most familiar in medical situations such as clinical trials. Chen states that efficacy evaluation is characterized by strong controls and a program that is implemented in a uniform way to a narrowly defined homogeneous target group (p. 200). He draws a distinction between such evaluations and what he calls effectiveness evaluation which assesses the effect of an intervention in real-world conditions.

Chen goes on to list the conditions favoring the use of effectiveness evaluation:

- There are questions about the feasibility of implementing a new program in the real world.
- There are concerns about the impact of a program, in the real world.
- Stakeholders require the evaluation to be relevant and of practical benefit to a program (p. 201).

In fact all of the above issues arose at one time or another in the PLTL evaluations. Professors adopting the workshops did not stop everything else and could not control for numerous other variables. They went on with the lectures, perhaps improving them. They continued to hold office hours. The workshop leaders varied according to their skill level and enthusiasm. In other words, they continued their work and implemented the workshops in the real world.

There is a question here about whether a highly controlled setting would have been desirable, even if it were possible. The reason is that, as Chen points out, stakeholders—in this case other instructors and anyone interested in the outcomes of the pilot workshops—would want to know whether the method is likely to work in their courses, and not simply in a controlled or ideal setting.

Instructors conducting the comparative studies did, however, establish the kinds of controls that were feasible, and that seemed most important. Almost always, the same professors taught the course for PLTL and non-PLTL groups; the same exams were used for divided classes; very similar exams were used for historic comparisons; and data were used to be sure that the groups were academically comparable.

One area that is often questioned with regard to controls is the impact of self selected groups in simultaneous studies. It may well be that students who self select into a two-hour weekly problem-solving session are more motivated than those who do not. The best response to this concern, we think, is to say that motivation is important but it is not out of one's control. Using the data, an instructor might say to a new class, "Students who were motivated enough to take these workshops performed better on tests than those who did not, so become more motivated and sign up for workshops." Nevertheless some people believe that the historic versus workshop data are more convincing because of this issue. (One PLTL study at Northwestern University, discussed in Chapter 7, found that those volunteering for workshops scored higher on an academic anxiety scale. Those conducting the study conjectured that this may mean that these students were less likely to succeed in the course—other things being equal.)

With this background we can move to the results of selected comparative studies. Since the initial implementation of PLTL more than 20 comparative studies have been undertaken. We will first present a summary of these studies, and then provide further detail.

In most studies, the basis for comparison was the percentage of students earning grades of A, B, or C. This standard was used because exams and grades are the currency of higher education, providing the best proof of successful academic outcomes. The average course grade generally rose with the introduction of PLTL, but not always by a great deal, and this was because as retention improved a larger number of struggling students were in fact passing with at least a C grade. In situations with initially higher percentages of quality grades, the average GPA increased significantly.

The comparisons in Table 2.4 show that, using non-PLTL as a base, PLTL groups generally out-performed non-PLTL groups, by an average of 10 to 20 percentage points, in earning grades of A, B, or C. Greater detail about these and other studies will provide a more complete picture of the performance-based evaluations.

Table 2.4 PLTL comparison studies: % ABC grades

Institution	Non-PLTL % ABC	PLTL % ABC
Historic Comparisons		
Rochester (Organic)	66 (n = 1,450)	79 (n = 1,554)
St. Xavier (Org & Bioch)	72 (n = 95)	84 (n = 116)
City College, NY (Gen C)	38 (three years)	58 (n = 484)
University of Portland (Gen C)	44 (%AB, three years)	73 (%AB, n = 99)
Prince George CC (A&P)	39 (three years)	53 (n = 34)
Miami, FL (Bio)	82 (n = 1,471)	85 (n = 1,584)
Evergreen CC (Gen C)	65 (n = 269)	74 (n = 74)
Randomly Assigned		
Pittsburgh (Gen C)	83 (n = 113)	90 (n = 130)
Self Selected Groups		
University of Rochester (Organic)	66 (n = 171)	79 (n = 119)
University of Kentucky (Gen C)	63 (n = 1,072)	84 (n = 92)
University of Ohio, Athens (Gen C)	77 (n = 292)	84 (n = 65)
Sierra College (Inorganic)	73 (n = 62)	94 (n = 82)
Portland State U (Organic)	74 (n = 119)	89 (n = 44)
Miami of Ohio (Gen C)	70 (n = 236)	75 (n = 116)
University of West GA (Gen C)	35 (n = 78)	49 (n = 145)
University of Puerto Rico Cayey (Gen C)	53.5 (n = 1,425)	69 (n = 424)

2.4.1 Grade and Retention Data

At the University of Rochester, after a one-year pilot, peer-led workshops were required for all students taking organic chemistry. Two-hour workshops took the place of one-hour recitations. With a long-time professor and careful records, Rochester did one of the more precise grade comparisons, determining that along with the percentage increase in quality grades from 66 to 79 percent, students taking workshops out-performed those without workshops by more than one-third of a grade, with a GPA difference of 0.4 (Tien et al., 2002).

Rochester attributed much of its success to a very careful leader training program (Tien et al., 2004). The heart of the program was a course developed and taught cooperatively by learning specialists and the professor: Vicki Roth, Lydia Tien, and Jack Kampmeier. The course employed concepts from several areas of learning theory combined with organic chemistry examples and problems. In an independent review of the Rochester data, Lyle and Robinson (2003) determined that the study clearly showed that PLTL improved student performance and increased retention.

One of the key concepts of the Rochester program was "cognitive apprenticeship," which was seen as a bridge linking classroom knowledge with pedagogical content knowledge. Using the approach, leaders were encouraged to consider how students would respond to a chemistry problem, and on the basis of this knowledge to discuss the best strategies to be used in problem solving. Problem solving itself was then explicitly explored using various approaches, and the leaders were encouraged to become role models.

At Portland State University, Professors Carl Wamser (2006) and Gwen Shusterman introduced Peer-Led Team Learning as optional accompaniments for the first two years of chemistry (general chemistry and organic chemistry) courses taught in large-lecture format. The peer-led workshops were offered as an additional one-credit, two-hour weekly course. About 30 percent of the students opted into the workshops. For organic chemistry, data were collected for five years, comparing the students who selected workshops with those who did not.

Student success (achievement of a grade of C- or higher) was compared for all those students who started each quarter. Student performance (percentage of total points accumulated) was compared for all those students who received a final grade at the end of each term. Student persistence (completing all three quarters successfully in the same academic year) was compared for all those students who started the fall quarter.

Students who elected to take the workshops had a slightly higher entering grade-point-average, based on all courses taken at PSU (3.27 vs. 3.14). Workshop groups out-performed the non-workshop groups by wide margins in all areas: ABC success rate (85 percent vs. 69 percent); performance (average course grades of 71 percent vs. 64 percent; course GPA of 2.90 vs. 2.51), and three-term persistence (57 percent vs. 28 percent); ACS scores (77th vs. 69th percentile). All of these are greater than can be statistically accounted for by the difference in GPA.

University of the Pacific. The correlation between workshop attendance and grades was dramatically demonstrated at the University of the Pacific in Stockton, California. Donald Wedegaertner collected the data in the table below (Table 2.5),

Table 2.5 University of the Pacific: workshop attendance and grades

Spring semester	Attendance: Poor	% ABC	Attendance: Good	% ABC
1997 (n = 37)	≤9/13 (n = 20)	38	≥10/13 (n = 16): 82	82
1998 (n = 19)	≤8/13 (n = 9)	67	≥9/13 (n = 10): 100	100
1999 (n = 38)	≤7/13 (n = 27)	45	≥8/13 (n = 27)	89
2000 (n = 17)	≤7/14 (n = 6)	33	≥8/14 (n = 11)	100
2001 (n = 34)	≤7/14 (n = 9)	67	≥8/14 (n = 25)	88

showing the percentage of ABC grades earned at different levels of attendance over a five year period. As mentioned above, poor attendance at workshops may be related to other factors that combine to cause poor performance.

University of West Georgia. At the University of West Georgia in Carrolton, Lucile Garmon tracked workshop attendance and grades for PLTL students in general chemistry even more closely with the results shown in Table 2.6.

Table 2.6 University of West Georgia: workshop attendance and grades

Attendance	A's	B's	C's	D's	F's	Ave
11 or 12 (n = 32)	6	20	4	2	0	83
9 or 10 (n = 27)	5	8	11	2	1	79
7 or 8 (n = 11)	0	1	4	5	1	68
Att. < 6 (n = 12)	0	0	6	4	2	64

University of Pittsburgh. In one of the few carefully controlled studies of PLTL in a lab setting, McCreary, Golde, and Koeske (2006) report on the results of a systematic comparison of conventional labs and peer-led workshop labs. In the workshop lab the TA was replaced by three or four peer leaders, each of whom worked with a group of about eight students. SAT scores and other measurements verified that the student groups were comparable. Tests were devised with 15 comparisons used to assess skills in four areas of higher order thinking related to the experiments. An independent grader was used. Students in the peer-led groups performed at a higher level in all of the comparisons, with nine being statistically significant ($p < 0.05$). In addition, students in the peer-led groups wrote at much greater length in response to items asking for explanations, and it was shown that quantity of writing held a positive correlation with quality. The PLTL faculty and researchers at Pittsburgh believe that the success of those in the peer-led groups was due to "much more interaction," in the workshops, "the weekly training sessions," and that the leaders, "were closer in age and experience to the students, and often adopted a less authoritative, more mentoring role (p. 810)."

2.4.2 *Standardized Test Results*

A number of PLTL faculty looked at the results of American Chemical Society standardized exams before and after introducing peer-led workshops. At Monroe Community College in Rochester, New York, John Cullen and Brian Edelbach

Table 2.7 Monroe CC: ACS organic II exam results

Year	National mean	Monroe mean	Difference	Percentile	N
1998	38.7	35	–3.7	40	26
1999	38.7	34.2	–4.5	36	30
2000	38.7	35.4	–3.3	40	36
2001	38.7	35.8	–2.9	43	33
2002[a]	38.7	39.1	0.4	54	19
2003[a]	43.1	42.4	–0.7	45	24
2004[a]	43.1	48.1	+5	62	36
2005[a]	39	52.2	+13.2	83	34
2006[a]	39	52.1	+13.1	83	42
2007[a]	39	52.4	+13.4	83	37

[a]Classes using PLTL

found the following outcomes for the organic chemistry exam for four years without PLTL and six years with PLTL. Table 2.7 speaks for itself revealing dramatic gains in student performance, and these increased over time as PLTL became rooted in the course.

2.4.3 *Comparisons with Other Interventions*

An interesting strategy for evaluating PLTL and comparing it with other changes was used by Joseph W. Wilson of the Chemistry Department at the University of Kentucky, Lexington. He kept track, over 10 semesters, of students from a lecture section of 150–200 students—in an organic chemistry course with a total enrollment of about 1,200 students—who selected one of several interventions. The department was not simply conducting an experiment but was trying different approaches based on resources and other priorities. The interventions were:

- Traditional recitations of 25 students led by a graduate student, meeting once a week for one hour. On average about 100 students per year selected this.
- PLTL workshop groups, meeting twice a week for two hours each time. On average about 25 students selected this option. They met as a group led by a graduate student with two undergraduate helpers. They broke into problem solving groups with about 8–10 in each.
- After three semesters, the PLTL groups described above were replaced by groups meeting once a week for two hours. These groups averaged seven students and were led by undergraduates. On average about 45 students chose this option.
- The remainder, a majority of the class, did not select any of the interventions and had the three lectures per week.

There were common exams and a common final. Those implementing the different options were particularly interested in two measures: first, academic success measured by grades, that is by the percentage of ABC grades as a portion of the total initial enrollment; and scores on the relevant ACS (American Chemical Society) exam. Records of entering GPAs were also kept. Results were as follows:

1. *Recitations*. For the three semesters of recitations, the recitation groups achieved success (ABC grades) at a rate of 10 percent higher than those in the class who were not in any groups, 67 percent vs. 57 percent.
2. *Workshops*. Both workshop groups achieved quality grades at substantially higher rates than the non-workshop group: 80 percent vs. 60 percent for the four hour, graduate-led workshops and 77 percent vs. 57 percent for the two-hour undergraduate-led workshops.

The data demonstrate that the PLTL groups achieved dramatic success compared with groups having recitations and those with no intervention. Eight years after this study, PLTL was still used at the university, listed in the catalog and considered a regular part of course offerings for general and organic chemistry. Dr. Wilson was retired but still supervising the workshops for organic chemistry.

2.5 Evaluation Data: Fidelity to the Model

The studies presented above demonstrate the positive effects of peer-led workshops on student academic performance in a variety of institutions. What was common in the implementation of PLTL at these sites was that the faculty adhered closely to the model that had been developed by the project and is described earlier in this chapter.

But there were courses in which the workshops did not enhance learning, and/or were not viewed favorably by students. In site visits, phone interviews, and other forms of data collection and analysis, the model of six critical components was useful in uncovering reasons for shortcomings in implementation and in student success. The critical components that form the PLTL model are presented here, with added detail about each component and examples of how incomplete implementation reduced the likelihood of success.

1. *Organizational arrangements*. Important workshop arrangements include appropriate space, time, and group size. The Workshop model recommends:
   - A two hour workshop, held once a week, with about six to eight students
   - That attendance be required
   - That space be adequate for concentrated small-group activities

At several sites, workshops were implemented with group sizes of about 20 students. We noted in observations that when a group of this size broke for cooperative activities, the leader had to float among as many as four or five groups. With each visit to a group, the leader had to take time and ask questions to find out where the discussion or problem solving was headed, whether it was misdirected, and what he or she could do to move the group into the right direction. This was a lot to ask of the peer leader and the problem solving process did not proceed as effectively as it should have. Further, it was almost impossible to work with the entire class as one problem-solving group. In the peer-led workshops, the peer leader is an essential ingredient as the group proceeds with its problem solving. Although the group of about eight sometimes breaks into smaller groups, this does not prevent the leader from keeping in touch with what each is doing.

The length of two hours, or at least 90 minutes, seems to be important in moving into a concentrated problem-solving mode. Initial time is always spent getting organized, answering questions, perhaps reviewing pre-workshop activities. Some workshops were implemented for 60 or 75 minutes, and these did not have the same positive impact on student performance as the longer workshops. Survey data about how workshop time is spent indicate that there is a correlation between the length of the workshop and the proportion of time spent on interactive group activities.

2. *Materials*. In general, professors adopting the workshop approach spent considerable time writing or adapting materials. The model recommends:

   - That materials be challenging and engaging but not so difficult as to discourage students
   - That the skills and knowledge developed in workshops be directly related to tests and grades
   - That the materials be suitable for small group work

Interviews with workshop professors revealed a general appreciation of published materials available through the PLTL project, combined with a need to adapt materials to local situations. The same materials have been reported to be too difficult at some sites and too easy at others. Materials prepared, piloted, and published by the Workshop Chemistry Project have been useful at a number of sites and may provide a model for other disciplines as they develop materials suitable for workshops.

At one site visit we found that students were not at all pleased with workshops, did not think that the workshops helped them with tests or that their grades improved. It turned out that the professor had developed very imaginative excursion-type materials that guided students in projects related to environmental issues. But these projects were not related to learning required for tests and so they did not result in improved student achievement or motivation.

At a few sites, the workshops were primarily devoted to a review of class-assigned textbook problems, not necessarily devised for group work. In these cases, there was less enthusiasm in the workshops which tended to become homework review sessions, or were like recitations in which the leader was consulted about answers.

3. *Training and activity of workshop leaders*. The training of leaders varies from site to site depending on the interest of the workshop professor, the presence of someone trained in science or math education, and the involvement of a learning specialist. The model recommends:

   - That workshop leaders be skilled in group work
   - That they be reasonably advanced in their knowledge of the discipline and as problem solvers
   - That they perform as facilitators rather than lecturers or TAs
   - That they have a training program before they begin
   - That they meet regularly, usually weekly with the course professor

The presence and activity of the peer leader distinguishes the workshop model from most other varieties of cooperative learning. Peer leaders generally understand that they are to guide rather than give direct instruction. But there is considerable variation in the style of problem solving that peer leaders adopt. Some take an algorithmic approach showing students how to set up and solve particular problems. Others are more conceptual, stressing the general principles that will enhance understanding and permit students to approach different categories of problems. At one site the workshop leader simply checked to see if students found the correct answers. Although this was still valuable as study time, the potential for improvement in problem-solving was minimized.

4. *Involvement of workshop professors*. Instructors adopting the Workshop model become involved in different ways with: the supervision of leaders; occasional observations of the workshops; and the development of materials. The workshop model recommends that the professor teaching the course:

   - Prepare, review, and update the workshop materials
   - Preview and model the workshop materials and activities with the student leaders
   - Be available to students and student leaders according to need

Turnover in faculty can have a serious impact on the PLTL project, and will be discussed at greater length in other parts of this study. By way of example, we note that at several sites when new faculty were hired either they themselves showed little enthusiasm for PLTL, or their department heads were reluctant to require responsibilities in this area, as they began their research and prepared their initial lectures. Not being involved with the workshop details, these faculty members were generally less convinced of the importance of workshops and students soon became aware of this attitude. By contrast, at several sites PLTL was adopted by an entire department, with notable success.

5. *Integration of workshops with the total course* As mentioned above, students taking courses that include workshops or who sign up for workshops as part of a course must see their participation in PLTL as an essential part of the course, as important as attending lectures and individual study. We have seen that most students do have a sense of responsibility about the workshops. It is also critical that faculty members integrate the workshops into the overall course structure. Site visits found that:

   - The workshops take a considerable amount of student time and energy. Consequently students must value the workshops or the impact will be considerably diminished.
   - Integration means that the workshop leaders are aware of the approach taken in the lectures and the professor's overall method.
   - The model requires that the professor refer to the workshops in lectures and at other times, indicating their importance to learning, perhaps using some cooperative work within the lecture.

At some sites, workshop details were coordinated by a faculty member not teaching the course or by a graduate student. In these cases student leaders were one or two steps removed from the lecturer, and the success of PLTL was partially compromised. When questions arose, students and student leaders sometimes lacked confidence that the priorities and methods of the instructor were clearly understood and interpreted correctly in the workshops. Similarly, when workshop leaders were graduate students, it was important that they meet regularly with the professor to align themselves with the teaching/learning strategies employed.

6. *Departmental and institutional support*. This is critical for institutionalization. The workshop program cannot survive without adequate resources, nor is it likely to survive if implemented by one or a very small number of faculty. A critical mass is required for the workshop approach to take root and become a normal part of the business of a department and institution. The model suggests that:

   - The workshop approach be extended across several courses and disciplines
   - That administrators such as department heads and deans support the program
   - That the institution provide local funding

Interviews with workshop faculty have uncovered a pattern regarding institutional support. At the outset, when first planning workshops, faculty members were generally enthusiastic about the pedagogical advantages, had acquired some resources with which to pay peer leaders and began to develop materials. Consequently they were not overly concerned with institutional support. But after a few years, particularly if colleagues began to adopt workshops, the need for on-site support became evident and even critical to insure long-term success. At sites where the department head did not offer support, the workshops were generally abandoned.

2.6 Workshops from the Leaders' Perspective

We have mentioned a number of times that the peer leaders were the key to successful workshops. They were also insightful and articulate in describing the workshop dynamics. At some sites they were asked to keep journals and entries were discussed in weekly meetings. These activities helped create a reflective atmosphere. Evaluation interviews with leaders provided rich data about the PLTL workshops. These interviews were recorded and transcribed. Excerpts are provided below.

How do you view your role as a workshop leader? *It's mostly guiding and helping them. That's what I do now. But when you first start doing it you accept the role of teacher. And finally you realize that's not your role. Instead of giving them the answer, I'll watch them do a problem and if they are going down a totally strange path, I'll say, "How about trying this?" And that has been better for the students. I have seen a change in grades. Last semester I saw a grade change from 40-something to 80, and from 60s to 90s. Just because students were working together, and also meeting outside of class.* (Peer leader for general chemistry in New York, NY, 1996).

What kind of changes have you noticed in the way students interact with one another? *I think in the beginning, the first couple of workshops, it took them a while to warm up, because they are used to competition, and they already know which ones are the brighter students and which are the weaker. And you have to coach them so that the better students back off a bit and let the quieter one speak up. They get a mutual respect for one another, and understand why they are doing what they're doing. At first they just look and if it's not in the book they say, "How do I do this?" And you have to tell them, "You have to think this through."* (Peer leader for general chemistry in Brooklyn, New York, 1996).

Did you have a training program at the start of the school year, if so what was it like? *The professor usually holds an orientation. When I began it was easier because there were only 10 in the group. He could give individual attention to each leader. Now it is more difficult. But we had an orientation where leaders like myself who have done it for awhile could give our experience. Then we broke into groups to work out some of the problems in the first set. That way we could see what exactly it feels like.* [Some of the leaders had not been in workshops as students.] *Then he asks: Did anyone feel awkward? Did anyone find it difficult to participate? That way we get a sense of how it feels to participate. There are always differences in interest levels; there's always a quieter one or a noisier one—even among the leaders. The leaders are supposedly more interested and want to do the problems. But we had to think about the students. Some of the students taking the course really don't want to be there; or don't want to be in the group. Dr. Wilson wanted to make sure that we could feel what it was like to be in a group. We want to be part of the group.* (Peer leader for organic chemistry in Kentucky, 2000).

What kinds of materials do you use? *I'm a little different, because I really get into this. I'll bring a model kit, or cut out different things from construction paper. We do skits. Like I'll get them to pretend they're molecules. It's really silly, but they get into it, and they're never going to forget how hydrogen bonding works. We're learning stochiometry this week and I'm bringing different kinds of candy to work on different kinds of groups. Anything to try make it more like real life.* [What got you doing this?] *Chemistry was really hard for me as a high school student and I did pretty poorly. So it is amazing that I have the major that I do. But I was frustrated and didn't enjoy it at all. So I came to college with a really bad attitude, and didn't really want to take the class that was required to go to medical school. But I was taking it and the professor starting doing all these experiments in class and so that kind of got me started thinking, 'Wow this is real; this is cool.' So I started looking for ways to make it interesting. And then I started to really like it. So I did a total 180 and began to love it. So then I would look for any way to make it more fun.*

[So the workshop was a great opportunity.] *Yes it really was. And I think maybe I am a little more compassionate than someone for whom chemistry comes easily. I know how hard it is. I think that a big part of the workshop is the discussion, like, 'Boy I had to work four hours trying to figure this out.' Because people think they are alone. So it is important to get students to understand the importance of group work.* (Peer leader for general chemistry at Miami, Ohio, 2000).

How is it working with students at different levels? *Some say they don't need it but when you see them there you see that they have weak points. Some say it so often that then they trap themselves and can't admit their weaknesses. Typically, they come from high schools where they have had chemistry and they do well for a few weeks and then they struggle. Others do well for the whole semester but they needed to interact with people more. They can be sub-leaders helping us explain and they can build friendships and confidence that help them as leaders.* (Student leader for general chemistry in Georgia, 2004).

There is abundant evidence that peer-led workshops can be of great benefit to students—improving academic performance, and helping freshmen get acclimated to college in a serious but friendly context. Chapter 6 is devoted to a study of the benefits of PLTL to the leaders.

2.7 Summary

Initial implementation and evaluation of Peer-led Team Learning led to the development of a model of six critical components. This model added clarity and energy to the program and to presentations for potential adopters. The critical components also provided a framework for monitoring implementation of PLTL.

When implemented according the model, PLTL has increased student engagement, motivation, and performance. Surveys and interviews with students and student leaders have uncovered a high level of satisfaction with the program. Studies comparing groups with and without workshops reveal that participation in workshops leads to a higher percentage of students earning grades of A, B, or C. Several studies have also documented a significant increase in performance on standardized tests.

When not implemented according to the model, PLTL invariably encounters problems. Each element of the model serves a purpose in contributing to the overall success. Site visits, interviews, and performance data have helped identify precisely where and why the program is not effective.

The peer leaders are central to the success of the program. A positive attitude, motivation, along with skills and continued guidance are essential if the leaders are to succeed in making learning more effective for students.

Chapter 3
Dissemination Strategies

Peer-Led Team Learning was implemented and evaluated at a diverse group of institutions. The founding group believed that the PLTL method would help students learn—regardless of the course level, the type of institution, or the students' background. They reasoned, based on their own experiences and the existing literature, that appropriately guided active learning would improve their courses. Implementation and early diffusion of the method was carried out by the founding group of faculty who published articles, developed a website and made presentations at local, regional and national chemistry meetings.

The success of Peer-Led Team Learning, both in terms of student performance and the enthusiasm of participating faculty, encouraged the founding members to take on the task of disseminating the PLTL pedagogy in a more structured way. As mentioned in the previous chapter, the effort was supported by two National Dissemination grants from the National Science Foundation (EHR/DUE). Simultaneously the PLTL project also participated in other projects committed to the improvement of science education for undergraduates, such as the Chautauqua Short Courses for College Teachers (http://www.chautauqua.pitt.edu/), Project Kaleidoscope (PKAL), and the Multi-Initiative Dissemination (MID) Project (http://www.cchem.berkeley.edu/~midp/index.html?main.html&1).

These PLTL activities were part of a larger philosophy that recognized the need of the participating faculty to be creative and to contribute to the national conversation about improving undergraduate education. Thus, the dissemination model outlined a practical approach for expanding PLTL within chemistry and introducing it to other disciplines. The model provided an opportunity for other faculty to contribute to the field of team learning and to exercise leadership. The plan built on success, with the conviction that faculty considering implementation of peer-led workshops in their own courses needed a thorough grounding in the method as well as support in implementation. A four-stage plan of dissemination was initiated.

L. Gafney, P. Varma-Nelson, *Peer-Led Team Learning: Evaluation, Dissemination, and Institutionalization of a College Level Initiative*,

3.1 Dissemination Model

The dissemination plan listed the following stages:

1. Stimulating interest
2. Creating a deeper understanding
3. Successful implementation
4. Developing scholarship and new leadership

For each of the stages there was a role to be played by the disseminators and by the target population of faculty and administrators. The dissemination team provided initial information; offered workshops with much greater detail on the method; then supported implementation and mentored new grant writing and local leadership. A unique aspect of the strategy was the involvement of students, the peer leaders, as partners in the development and dissemination of the program.

3.1.1 Stimulating Interest

In the first stage of dissemination, experienced and successful PLTL faculty members and students made the method known to colleagues within their own and neighboring institutions, and through other professional contacts. This sowing of the seed was done in a variety of formal and informal ways: in the many brief conversations that professors and students have with one another; through talks and panels at regional and national conferences; through the PLTL national organization, newsletter and website; through presentations at faculty development conferences; and in publications about related research and the evaluation of PLTL.

As part of the evaluation of informational conferences disseminating the four pedagogies (Multi-Initiative Dissemination) mentioned above, a survey item asked about the "barriers to your adapting or adopting what we have been discussing." The following selected responses demonstrate that participants were quick to identify potential issues, problems, and barriers to successful implementation of PLTL or other similar initiatives. Participants listed the following concerns.

- Money. Training of TAs or peer leaders.
- Multiple sections taught by multiple instructors.
- Time to prepare materials
- Getting departmental support (but that also is an opportunity).
- It is my first year at school; not tenured and not certain of support.
- Putting all your eggs in one of these baskets is no great improvement over doing just lecture—if you do them well.
- No existing time period (recitation) or student TA force to help implement. Will have to be very creative.
- Administrative hesitation and lack of support; this is one more thing to do above and beyond what I currently do.
- There are many new programs; we must decide on one and go with it.

- I don't teach the same classes several years in a row, so time invested will only pay off if others who teach course collaborate and agree to use same approach.
- My own time budget. To add to time for this class modification is more likely to be gradual than a giant leap.

Most of the issues identified are substantial. The faculty members participating in the informational workshops had an accurate understanding that initiating workshops and other programs would take time, funding, organizational activities, and departmental approval. Some were not sure that overcoming all the barriers would result in improved student learning. But for many the challenge of better teaching and learning was an overriding concern. So they solved the problems and overcame the barriers to see for themselves whether PLTL workshops would take students to a new level of learning.

A few anecdotes can provide the flavor of the initial information flow about PLTL. These are taken from the scores of faculty members who learned about the program and implemented workshops.

- A chemistry professor at a New England state college learned of the first NSF chemistry grants, requested more information, thought the workshops would be beneficial particularly with large classes, applied for and received a PLTL minigrant (Workshop Project Associates).
- A professor at a mid-western state institution who attended a Lilly-sponsored meeting on teaching and curriculum and was impressed with the excitement of PLTL students, went to a Chautauqua short course to gain a deeper understanding of the program, and then implemented the method.
- A professor from a community college in New York State who had been using small group work for case studies in biology, learned of PLTL from a colleague, tried it and was pleased with the results.
- A professor from a research university had learned of the method but thought it was for remedial instruction. When he found that an institution similar to his was using it with all students, he attended a national PLTL meeting organized by the project and then introduced the method with his general chemistry honors class.

So we can see that faculty members gained initial familiarity with PLTL in a variety of settings and were attracted to it for different reasons. As was mentioned earlier, it was generally those who had already introduced new teaching/learning practices who were most favorably disposed to PLTL. Berke (2003) had implemented a number of active-learning curricular changes when he learned of and almost immediately became enthusiastic about PLTL. Having someone from a similar institution successfully implement a pedagogy also seems to be important to faculty before they were willing to invest time in a new approach.

3.1.2 *Creating a Deeper Understanding*

In the second stage of dissemination, faculty who had some initial familiarity with the method attended workshops, generally of two or three days in length. Some of these workshops were sponsored by educational agencies such as the Chautauqua short course for college teachers; others were hosted by the PLTL project. These workshops required a substantial commitment of time and work and a majority of the participants were seriously considering PLTL.

These training workshops focused primarily on an exploration of the six critical components of the workshop model with emphasis on leader training and preparation of workshop materials, with additional information about evaluation and suggestions for startup funding. Student leaders from an institution using PLTL would regularly participate in these presentations, and they were often the most convincing part of the program because they were poised and enthusiastic about the many benefits of PLTL.

The importance of the peer leader participation in dissemination cannot be over estimated. They generally provided a demonstration of their own role by leading a workshop while faculty participants played the role of students. They also explained the benefits of the method based on their experiences, answered questions, and in general were articulate advocates for the model.

Professors accustomed to more passive students were deeply impressed. Comments such as, "The most effective part was the example" (Miami, 2000 PLTL conference), were common. Faculty members saw, sometimes for the first time how empowering the method is for students to participate in the design of a learning environment. For the most part, these were not students from highly selective universities. But they demonstrated skills that would make them competitive applicants for graduate school and would be valued by employers. Chapter 6 provides a detailed discussion of the PLTL benefits for leaders.

During the spring and summer of the dissemination grant's first year, two and three day PLTL workshops were held at five sites: Miami, Florida; Portland, Oregon; Pasadena, California; Philadelphia, Pennsylvania; and San Juan, Puerto Rico. We have survey forms from 82 participants (Gafney, 2000b).

Participants at these conferences, unlike those for the brief informational sessions, were almost all seriously considering adopting PLTL. Their concerns were about implementation issues and are shown in the following table (Table 3.1), in absolute numbers (N = 82). Some participants listed more than one concern.

In spite of their concerns about issues in implementation, respondents gave high marks, on a 1–5 scale, to the value of topics and activities covered at the conferences. The following table (Table 3.2) presents the means of these ratings.

Table 3.1 Faculty concerns in implementing change

Concern	(N = 82)
Support: institutional, departmental, collegial	27
Cost/funding	23
Scheduling; finding space	15
Time for implementation activities	14
Recruiting and training peer leaders	11
Developing materials	10
Pedagogical fit; student interest	8

Table 3.2 Ratings of PLTL topics at dissemination conferences

Topic	Average Rating
PLTL model	4.5
Leader training	4.5
Materials	4.2
Evaluation	3.8
Mini-grant (WPA) program	4.4

These conferences and many others like them achieved the goal of motivating and preparing participants to implement PLTL, with appropriate support. More than 90 percent of participants thought that implementation was feasible for them.

3.1.3 *Successful Implementation*

The third stage of the dissemination plan was actual implementation at a new site. This was often done on a pilot basis, sometimes with only a fraction of a class participating. A key part of this third phase was the support provided by experienced PLTL practitioners to those beginning to use the method. Issues often arose, for example, about training leaders, adapting materials, and overall organization. Issues in these areas could be daunting. But the experience of those who had successfully implemented the program provided helpful guidance to those initiating PLTL.

During the years of NSF funding, initial implementation of PLTL often took place with financial assistance from the project-sponsored mini-grants. Requirements for proposals and reports were prepared and administered in large measure by Varma-Nelson. These mini-grants not only permitted instructors and institutions to implement PLTL but also created a wealth of useful data about issues surrounding new implementations and dissemination. Chapter 4 is devoted to this aspect of the project.

For those implementing PLTL in general chemistry, organic chemistry, and a combined general, organic, and biological chemistry (GOB) course, the workshop materials developed and published through the project facilitated startup of PLTL. Many professors adapted the materials to suit their needs, but having them made the initial planning far easier.

The project also published a manual covering all aspects of the project and a peer leader handbook. These were reference guides, following up on the training sessions and permitting more confident implementation. Materials for chemistry were published and two were later revised as second editions (Gosser et al., 2006; Kampmeier et al., 2006; Varma-Nelson & Cracolice, 2001). Since 2005 a set of carefully developed modules has also been available on line for implementers in biology (Griswold et al., 2005). Materials for math and computer science courses have also been developed by individual implementers.

3.1.4 Developing Scholarship and Leadership

The fourth stage in dissemination as planned was that the new implementers would take on leadership roles, promoting the program on their own campuses and beyond, contributing to the evolution and growth of the PLTL model. This often happened and so deeper involvement grew, not only of PLTL users but also of experienced practitioners who continued to tell the story of the method and its benefits. Further detail on this aspect of program growth is contained in the following chapter.

As an aid to dissemination the PLTL website contains a page with 10 items that can serve as a starting point for those disseminating information about the program or to those who want to receive such information (Varma-Nelson & Dreyfuss, 2003). Materials include: a sample brochure to be downloaded or used as a template to promote the program; PowerPoint presentations explaining the method; a listing of more than 240 presentations on the method; a page with links to a number of sites using PLTL; guidelines for applying for small grants; sample workshops; and a checklist of issues to be considered before taking student leaders to conferences.

These resources are intended to assist new PLTL implementers to prepare and present the method at local or regional conferences and workshops. They are not exhaustive or arranged in a way that would obviate the need for individual work. In fact it is important that as individuals disseminate PLTL they put their own stamp on the message, relating in a personal way how and why the method works for them. But these web-based resources decreased the time needed to prepare dissemination presentations.

Other activities carried out by this second-wave of PLTL faculty included: several very careful studies of student performance and student leader experiences that resulted in publications; promotion of the program outside the United States, with several trips to China, India, Cuba, and Europe; use of the program in connection with grants awarded by NIH, FIPSE, and others; numerous presentations at regional and national conferences; successful PLTL programs established in mathematics and biology. The PLTL website lists many of the presentations, workshops, and articles related to the project. Some of this information is also listed on individual faculty and university websites.

The data indicate that the dissemination plan worked quite well. A large number of faculty, potential student leaders and learning specialists attended presentations and workshops, and the program of mini-grants catalyzed PLTL implementation at over 90 new sites. Evaluation has found that implementation at new sites was mixed both in the quantity and quality of activity.

3.2 Using Diffusion Practices to Improve Education

There are diffusion and dissemination lessons to be learned from other fields. Dearing (2004) points out that the lack of diffusion of best practices is a critical issue in health care. He finds that new and local solutions are continually rewarded, to the neglect of programs with proven effectiveness. We propose that the same holds true in education, and concur with his statement that, "there is little if any correlation between program effectiveness and program persistence" (p. 23).

The stages in dissemination outlined above, and employed by the PLTL practitioners, offered a strategy for making the method known, and helping faculty get started. The stages used by Rogers (2003) to explain the innovation-decision process cover the same ground in greater detail and can be used to throw additional light on what was happening as the ideas and methods of PLTL began to spread. Rogers' theory and categories will be used in the following analysis of PLTL dissemination. We use "dissemination," the more common term currently used in educational practices and "diffusion," interchangeably.

3.2.1 Knowledge

The first stage Rogers discusses is knowledge. He considers three types of knowledge in response to three questions, all of which are pertinent to PLTL. The first question is, "What is the innovation?" He calls the answer to the question, awareness-knowledge and he believes that those promoting an educational innovation or reform effort should be able to say what it is in a few words. PLTL practitioners have had many encounters in which colleagues or faculty want this very basic information. The answers are straightforward, "It's about small group learning, with a trained peer leader, and carefully developed materials; it's part of a course."

The second knowledge question is, "How does it work?" Answering this takes more time, and involves a full exposition of the model with its six critical components, providing what Rogers calls how-to knowledge. He makes the telling comment, well known to PLTL advocates, that, "when an adequate level of how-to knowledge is not obtained prior to the trial and adoption of an innovation, rejection and discontinuance are likely to result" (p. 173). The reason is that without clarity about how to implement, there is an increased likelihood of mistakes, confusion, and unproductive outcomes. He finds that more complex innovations require a higher level of how-to knowledge than those that are less complex.

The implementation of PLTL is more complex than it first appears. Adding problem-solving and concept-building workshops to a course sounds fairly simple but this is not a small change at a university or college. There are a number of issues that may create problems if not anticipated or handled promptly and appropriately when they occur. We mention a few of the more commonly encountered issues.

- Finding suitable space for the discussion and problem-solving that takes place within the workshops. We have observed workshops in lecture halls, classrooms, labs, seminar rooms, student lounges, and cafeterias.
- Scheduling and taking attendance.
- Making sure that materials are at the right level and suited to the allotted time.
- Selecting, training, and supervising leaders.
- Keeping the workshops closely related to lectures.

It has become clear that failure to anticipate these issues can derail a project before it has had a chance to succeed. If the problems listed above, and others, arise as surprises then implementation can be quite bumpy, a high level of student dissatisfaction is likely, and one may question whether the project is worth the effort. Similarly, if PLTL is viewed as nothing more than a weekly study session with a guide, then it is likely that many students will not attend and those who do will not reap significant benefits. A reasonably high level of how-to knowledge will not eliminate all implementation problems but it will help faculty anticipate them. As the saying goes with regard to military operations, one must plan in great detail, knowing that once the battle is engaged much will not go according to plan.

The third knowledge question is, “Why does it work?” The answer requires an understanding of principles. This is an interesting area with regard to PLTL, and there seem to be a variety of answers related to different theoretical frameworks. Some point to constructivist theory that emphasizes the active participation of the learner, enhanced by cooperative activities in a social setting. Others believe it is simply a matter of getting students together for structured and guided problem-solving. Others have identified the coordination of reciprocal teaching and studio instruction. Still others see the intervention and modeling by the peer leader as the critical ingredient. Some stress the power of friendly feedback. And some who are enthusiastic about the method do not seem to worry about why it works. We believe that more research needs to be done to learn why the PLTL workshop environment helps students become more successful.

3.2.2 *Persuasion*

After knowledge, the second stage in the Rogers model is persuasion. The individual must become “more psychologically involved with the innovation.” The potential PLTL adopters, in this case college instructors, worked through uncertainties associated with the method, asking about the advantages and disadvantages, not in the abstract, but in their own settings, and in conjunction with their own priorities. These local evaluation data were persuasive. The outcome at this stage is a favorable or unfavorable attitude about the innovation. Generally a favorable attitude will result in adoption, at least on a trial basis. But there can be a gap between attitude and action.

In the case of PLTL we can conjecture about why some instructors were quickly persuaded and others were not. Those instructors who found PLTL suited to their needs were the ones who had been actively looking for new ways to engage students before they found PLTL They recognized teaching/learning issues, and were searching for solutions.

- A chemistry instructor at a southern state institution had been using tutors and had looked at Supplemental Instruction when she attended a Chautauqua conference on PLTL and found that it answered a number of questions about improved teaching and learning.

- A professor at a mid-Atlantic school was hired in biology education. She observed PLTL as used in the chemistry department and adapted the method to biology.
- Several instructors had been considering other approaches to group study and tutorial support, but recognized advantages in PLTL—particularly in the leaders' role and the integrated activity of the instructor—elements that made them select PLTL.

3.2.3 Diffusion Decisions

Rogers points out that, before making a final decision, most adopters want to try an innovation first on a trial basis, "to determine its usefulness in their own situations" (p. 177). He further notes that a cultural context or the climate of an institution influences the likelihood of adoption. For example, in some situations there may be strong group pressure in decision making; in others a more individualistic atmosphere may prevail.

Dearing (2004) provides useful ideas about diffusion that to some extent bridge Rogers's notions of knowledge and persuasion. He found that decisions about adopting an innovation are guided by: what people think about it; what they think others think about it; and how they compare it with other innovations (p. 26).

This "what others think about it," was operative in the case of PLTL. For a variety of reasons, faculty members are more likely to listen to colleagues from disciplines and institutions like their own (Foertsch et al., 1997) and to be in more frequent contact with those from institutions in the same geographic region. For this reason a supplement to the PLTL national dissemination grant was directed toward dissemination by community college faculty to others at community colleges. Also, consortia of colleges sharing ideas, leader training, and materials existed in the New York City area, around Rochester, New York, in the Midwest (Chicago and Indiana) and in Portland, Oregon among other locations.

In Portland, the Oregon Collaborative for Excellence in the Preparation of Teachers (OCEPT), an NSF-funded initiative to improve the quantity and quality of science and math teachers, had assembled faculty members from all levels of higher education to consider new approaches to science teaching. This consortium provided fertile ground for Oregon faculty to learn about and implement PLTL. As an unintended consequence, PLTL also offered an opportunity to increase the number of science teachers in the state, as undergraduate peer leaders developed an interest in teaching.

The implementation stage occurs when the innovation is adopted. Rogers discusses the "unit of adoption," which might be an institution, a department, or an individual. He finds that more complex issues arise when the implementer is an organization rather than an individual. With regard to PLTL, the adopter was generally an individual or small group, but the organization played a critical role in implementation since the institution provided funds, space, time, and legitimacy. Rogers finds that re-invention—the changes in the idea or practice as it moves from adopter to adopter—is a natural process and not necessarily bad. We will return to this in Chapter 4, finding a wide range of adaption strategies, some of which were more successful than others.

3.3 Attributes of Innovations

A second framework of Rogers, useful in the analysis of PLTL dissemination, relates to attributes or qualities perceived by potential adopters. We will discuss four of these attributes, and their relevance for PLTL.

3.3.1 Relative Advantage

Potential adopters want to know that an innovation will be worth the cost, in terms of time, effort, and money. After a PLTL presentation at a research university, several faculty members commented that they believed the workshops were good for student learning but implementation activities were too demanding. Instructors are faced with responsibilities in several areas—teaching, conducting research, serving on committees, publishing and advising, and more. Decisions about time-consuming educational innovations are made in this context. Will the new approach support or perhaps distract from other responsibilities? Incentives and rewards can play a role by increasing the relative advantage or reducing the costs of implementation.

3.3.2 Compatibility

Closely related to relative advantage is compatibility—not only with professional activities but also with the values and beliefs that affect an instructor's behavior. Those who were already trying approaches to active learning and cooperative work were quick to notice the strengths of PLTL. But the model was sometimes viewed as incompatible with a commitment to traditional lectures, individual study, and interest in a competitive rather than a collaborative academic environment. Some departments thought that the use of undergraduates as peer leaders was in conflict with time needed for them to engage in research and/or with a commitment to have graduate teaching assistants conduct recitations. But the opposite was also true. Faculty at research universities found the give and take of workshop discussions was good preparation for participation in discussions about research findings.

3.3.3 Complexity

Rogers finds that complexity is negatively related to the rate of adoption, and once again his insights are consistent with experiences in the dissemination of PLTL.

Some people thought the model was too complicated when they realized how much detailed work was required to select and train leaders, develop new materials compatible with the lectures and suited to group work, find meeting space, and convince students that this extra effort was really valuable to them. One professor adopted the method with a relatively small honor's class, saying that implementing with the entire introductory class of 250 students would be, "like planning the Normandy invasion." A graduate student at a state university mentioned that there were about 200 students in each introductory chemistry section, and there were 10 of these sections. Clearly, complexity would be an issue in implementing PLTL at such an institution. But the large classes gained significant benefits once the complexity issues were addressed, and a number of institutions did implement PLTL with large introductory classes.

3.3.4 Trialability

This quality refers to the degree to which an innovation can be tried experimentally. In innovations generally, trialability seems more important among early adopters. This was partially the case with PLTL. After the program was well defined, adopters had better knowledge about what to expect. But college professors are independent minded and trialability remained an issue as each adopter wondered if the method would work well in his or her situation.

A successful strategy, used by several professors was to limit the trial to a subsection of a large class and implement adhering to the critical components before scaling up. Adopters sometimes diluted the method in order to try it, for example by limiting the time spent supervising student leaders or by simply using materials from the text book. They then found that the program did not have the success that was anticipated or students were not as enthusiastic as they had hoped.

3.4 Additional Issues in Dissemination

We have touched on a number of areas in which the PLTL model was viewed as in conflict with individual or departmental priorities. We will now consider these issues in greater detail.

3.4.1 The Role of Peer Leaders and of Faculty

A central belief in the implementation of Peer-Led Team Learning workshops is that peers, only a year or so older than the students in a course—perhaps a difficult science course—can play an important positive role in the teaching/learning process.

This role is not something that is handed to the undergraduate leaders by default, because there are not enough professors or graduate students. Rather, the undergraduate is seen as a recent student of the material who remembers well the issues faced by new learners. The peer leader can therefore assist students in problem solving and in their conceptual development, in ways that the professor or even a graduate student cannot.

For many this is a difficult bridge to cross. Those of us involved with PLTL have seen and heard professors express reluctance and anxiety at the mistakes that the peer leaders might make when explaining material to other students. Saying that they are facilitators, not teachers, does little to alleviate these feelings. Fortunately, faculty members with these anxieties are few in number compared with the many who find the use of leaders to be a positive addition to the teaching and learning process.

Finkel and Monk (1997) articulate the self-image of many faculty members. College professors are not only teachers, they are experts by profession, seeing themselves as representatives of their fields. As such they, "are bound to feel a strong personal discomfort in the presence of the kinds of imprecise, loosely connected, unintegrated comprehension that students have of their subjects" (p. 7). This is an insightful comment that most of us would find rings true. No matter what the discipline, professors feel compelled to correct their students, not only so that the student will learn but also to set the record straight, so that a misconception or error will not infect the topic.

Therefore, for the faculty member contemplating PLTL, the question of using undergraduate facilitators can be emotionally charged: "Will wrong information be put forward, and will this be done in my course, in my name?" Finkel and Monk believe that the problem can be solved by shifting the consideration from roles to activities. If the teaching/learning process is defined by roles, then the teacher's role is that of professional expert who knows the discipline and knows best how to transmit knowledge. Students are dependent and their role is to accept the knowledge.

But if we define the process in terms of activities, then a shift in thinking can take place. The instructor may at times be the master, directing and giving knowledge to the students. But at other times, the instructor may move to the sidelines while students become more independent—working with one another, learning together, creating knowledge cooperatively, as they learn by interacting with one another. The instructor may put aside his or her anxiety and even feel some relief as a helper rather than as the indispensable expert. A PLTL professor eloquently expressed the positive results of moving to the sidelines, as follows.

> *The biggest impact is something that I have discussed at length at several meetings and workshop conferences. I don't know how it would affect their mark at the end of the semester but it certainly makes them more open and inquiring and more ready to challenge than they usually would be in my presence alone. The student leaders, for instance, are regarded by the people in the course as almost peers. So there is greater freedom to challenge what they say and to ask them questions. I've watched this from the background. It's one of the things that I noticed right away—so that by the end of the term they challenge me more. I can't prove it, but I've seen people who are shy and reticent in class begin to ask questions of me and actively begin to explore. I see that as the most positive thing. We have a community college here. Our students come in with a bad self image. If at the end of the semester they can open their minds up, they can become a little more assertive—it's a fabulous*

> *change. It really is and it's almost secondary to see their marks go up.* (Community College professor)

For professors who want to maintain the guardian-of-knowledge role, the "unsupervised" workshop group may be difficult to tolerate. But in our experience, faculty with this mindset were not large in number. Professors who successfully implemented PLTL saw the workshop as a productive group problem-solving and concept-building activity, and they were able to relax with the knowledge that students are making progress not only in learning science, but also in doing science and in deepening their conceptual understanding.

In considering the concerns mentioned, some sites encouraged student leaders to email questions to the professor. This proved to be a practical way of maintaining contact and reducing the spread of misinformation. Finally, some PLTL adopters have noted that the presence of potential local leaders at the introductory training sessions for faculty helped alleviate faculty reservations about the program.

3.4.2 Diffusion by Grant and Cluster

Larson and Dearing (2002) build on the work of Rogers particularly in pointing to the possibilities of using granting agencies to advance the diffusion of innovations and by using "cluster diffusion." They identify a key issue in the fact that there are a multitude of innovations with little diffusion. They point out that, particularly in the field of technology, innovations of proven effectiveness frequently are not adopted, while less effective methods remain is use. In addition, government agencies and foundations do not offer funds or incentives for diffusion to match the resources they put into innovation.

Larson and Dearing find that the study of diffusion has developed a solid and well tested theory that can be used to explain the spread of innovations—as we have done for PLTL above. They stress the importance of opinion leaders and the fact that their presence can make diffusion easier and more efficient. They use the term "cluster diffusion," to describe a process through which several related concepts and practices can be promoted together. The authors then make the following point that has been particularly true about educational innovations and that is worth considering in some detail.

> *The special promise of a complementary cluster diffusion strategy is that when people adopt one innovation they frequently are amenable to adopting one or more related innovations, opening a wide window of opportunity for change agencies. The special promise of a competitive cluster diffusion strategy is that change agencies such as private foundations do not have to "pick winners" and run the risk of seemingly advocating one intervention at the expense of others. Rather, potential adopters can choose from among a set of competing innovations to solve a given problem.*

College professors and instructors fit the model described above in that those who are interested in fresh approaches to pedagogy and curriculum are often willing to consider more than one new idea or practice. PLTL recognized this and was involved in several projects disseminating a variety of innovations, as mentioned at the beginning of the chapter.

An issue in adapting or combining methods is that important features of a program or pedagogy may be lost and effectiveness compromised. Consequently, those disseminating PLTL have mixed feelings about adopting the cluster model. The PLTL approach has been implemented with success and verifiable benefits to students in a variety of disciplines, courses, and institutions, and with all kinds of students. In addition, the method can be used in conjunction with other new approaches such as calibrated peer review (University of New Hampshire); problem-based learning (University of Rochester); personal response systems (IUPUI); and POGIL (University of South Florida). But if the critical components for success are not preserved, there are always significant weaknesses, and benefits to students are diminished.

3.4.3 Opinion Leaders

There are a variety of leadership roles in higher education. College presidents, deans, and department heads are leaders by election or appointment and they have clearly defined roles. In research, leadership is earned through exploration, discovery, and publication. With regard to teaching, some faculty members are recognized as being particularly successful. This recognition leads to influence as others adopt the approaches they admire. But real leadership in teaching emerges when exceptional instructors extend their influence by describing what it is that makes them successful and by defining practices that others can adopt.

Larson and Meyer (2007) provide a vignette describing a young faculty member who acquires ideas about teaching and learning at conferences and workshops. This is essentially one-way communication. After some time this person talks to a colleague whom he respects within the department, and this begins a two-way communication that leads to further interest and experimentation. In the proper climate, these experiences may create pedagogical change. The article then describes the use of a survey asking about opinion leaders. Results of the survey revealed that opinion leaders were rather narrowly focused within the discipline, often in the department and the sub-specialty of the respondent. Faculty respondents described opinion leaders, "more on research experience than on teaching competence."

Within an institution the chair of a department often carries significant influence. This is particularly true if he or she has reshaped the department or come into office at a time of transition. At several institutions, the chair of chemistry or biology decided to initiate PLTL in all the introductory courses, found the funding to do so, and led the participating faculty in developing a plan for PLTL. At several other sites the chair noticed the success of PLTL and provided strong support, encouraging others to participate. At one other institution, the chair permitted limited implementation of PLTL, discouraging its use beyond general chemistry. From the perspective of dissemination, the lesson is that the leadership or support from the department chair can be highly significant in successful implementation.

It is well established that positions of leadership among college level faculty are discipline based (Foertsch et al., 1997). The PLTL founding group and subsequent disseminators recognized this fact and used it in dissemination. While faculty members sometimes attracted those in other disciplines to adopt the method, this was almost always among colleagues within an institution. Real progress across institutions came about only when new leaders had emerged for the discipline.

There was in the diffusion of PLTL a serendipitous quality that does not easily submit to analysis. One of the most successful participants, in his classroom and in local and national dissemination, describes how he heard about PLTL from a previously unknown faculty member from another institution, and immediately signed on as a pilot site for the program. More study is needed about the role and influence of opinion leaders in the dissemination of PLTL and other programs. At this point we would say that opinion leaders did not play a dominant role in the dissemination of the program. It was the model itself, as presented by faculty and peer leaders, that attracted attention and responded to a felt need among faculty.

3.5 PLTL Leadership and Coordination

Coordination of the initial implementation and later dissemination of PLTL was determined by the NSF grants. The emergence of new participants and leaders of the program also grew out of these grants. The founding group of faculty were tied to the project by their commitment to improving PLTL at their own sites, by funding, and by their individual responsibilities within the project.

Responsibilities within the project varied and the principal participants adopted somewhat different roles. These roles were more a matter of emphasis than exclusive or proprietary concerns. The roles and responsibilities included:

- Offering dissemination presentations and workshops
- Research on academic performance of comparative groups.
- Research on the pedagogical foundations of PLTL.
- Developing materials for particular courses.
- Conceptualizing the PLTL approach in disciplines beyond chemistry.
- Investigating and responding to the needs of community colleges.
- Organizing conferences and publishing the project newsletter.
- Coordinating the overall efforts of the project team.
- Managing the mini-grant program.

The distribution of responsibilities and tasks was generally beneficial to the project. Specialization is key to the success of any organization. Specialization in the PLTL project took place naturally as those with experience or interest in a particular area began to work in that area. Regular meetings and informal contacts proved useful to maintain the flow of information. The grants with visiting committees provided a level of accountability, as did the informal exchanges and "peer reviews" that were a natural part of the programs development.

Chapter 4
Adopting and Adapting PLTL: Successes and Limits

In Chapter 3 we discussed the strategies of dissemination employed by the PLTL project and analyzed these in the light of dissemination theories, particularly those of Dearing and Rogers. We saw that PLTL adopters and potential adopters generally worked through several stages. First, those with an interest in team learning needed more detailed knowledge about what PLTL is, how it works, and why it is effective. Second, potential adopters needed to believe that in their particular circumstances they could effectively implement the workshops. Then they would try the workshops, often on a pilot basis, in order to satisfy themselves that the program could be reproduced in their own environments and that it would promote student success and engagement at a cost that they were willing to pay—in terms of time, money, and commitment.

Rogers also found that adopters of a new practice generally adapted the methods in ways that they believed were compatible with their circumstances, priorities, and resources. This chapter will present the experiences of new adopters first as captured in reports submitted after an institution had experimented with PLTL, and will then provide added detail obtained through site visits and phone interviews. These visits and interviews made contact with sites that were of particular interest because of the adaptations made, local issues, or problems encountered. Through this analysis we were able to identify conditions that lead to successful implementation, and the issues that militate against successful adoption of the PLTL method—with lessons transferable to other initiatives.

4.1 Workshop Project Associate (WPA) Mini-Grants: Scope of the Program

One of the key activities of the two national dissemination grants funded by NSF was the mini-grant initiative, known as the Workshop Project Associates (WPA) program. This activity fits clearly under the third stage of the PLTL dissemination model, providing funds and technical expertise to adopters. Funds were used primarily to compensate the workshop leaders. The payments, although

L. Gafney, P. Varma-Nelson, *Peer-Led Team Learning: Evaluation, Dissemination, and Institutionalization of a College Level Initiative*,

modest, emphasized the importance of the peer leaders to the PLTL model, and helped secure a strong commitment from them. After PLTL had proven itself, funding for the leaders was usually provided by the department or some other institutional source.

The mini-grants served another important function. Potential adopters were assured that compatibility with local needs was essential, and experimentation was expected. WPA grant recipients were welcomed as colleagues in the program. It was assumed that their experiences would provide new insights and would strengthen the project in both practice and theory. Experienced PLTL practitioners entered into partnership with the grantees so that the support offered became a two-way street, with new strategies and insights flowing from as well as to the new adopters.

It has been our experience that demonstrated success of PLTL on a campus is particularly convincing to administrators who provide the budget to sustain the program. The process also facilitates the careful adoption of PLTL through experimentation, data collection, and modifications to address the needs of each student population.

A large part of the WPA program was administered under the direction of one of the dissemination grant's principal investigators and co-author of this report, Pratibha Varma-Nelson, formerly at St. Xavier University in Chicago and then at Northeastern Illinois University.

Varma-Nelson, in consultation with others in the project, developed a set of guidelines for processing applications and administering the WPA one-year mini-grants to initiate PLTL. These guidelines included:

- *Preparation of an outline describing what was expected.* Proposals were to include a description of the course, materials to be used, strategies for training leaders, and a plan showing how the PLTL model would be followed. The proposal process was intended to verify that the faculty member submitting it had the basic "how-to" knowledge needed to adopt and adapt the program.
- *Review and rating of the proposals by at least three members of the PLTL team.* Experienced PLTL practitioners rated the overall quality of the proposals, considered the likelihood of success, and made recommendations for funding, modifications, or re-submission.
- *Matching funds from the institution in real dollars, not in-kind services.* Funds were used primarily to provide stipends for the peer leaders and sometimes for faculty travel. Matching funds were required to ensure commitment from the institution, and also to foster discussions on campus about the need to introduce a line in departmental budgets for the peer leader salaries. Matching funds also resulted in administrative oversight, and this in turn made likely the program review required for institutionalization.
- *Preparation and submission of a brief report after completion of the one-year grant period.* A format for this report was provided to the grantee. The report was to provide quantitative and qualitative data about the numbers and experiences of students and faculty involved, and also about expectations and issues in continuing the program.

Although these grants provided a maximum of only $5,000 per course, or $10,000 per discipline for two or more courses, there were a large number of applicants. In the five years of the program, 92 WPA grants were awarded, selected from approximately 110 applications, totaling $491,866.40 and matched by $544,443.34 in institutional funding.

The 84-percent rate for funding applicants was high compared to a typical program offering grants. This is because the number of potential applicants had been obtained from a much larger pool through a self-selecting process. Faculty members attended professional development seminars about PLTL. Those with a serious interest then attended two- and three-day workshops. A number of these individuals consulted with members of the founding group. Consequently, of those expressing initial interest in the program it was a minority that finally decided to apply for WPA grants in order to pilot PLTL. In view of this process, it is not surprising that a high percentage of proposals were funded.

Preparation of WPA proposals and implementation of workshops included some unexpected issues. Some of the applicants were writing grants for the first time. As such, they needed guidance but also acquired skills for future use. Both applicants and those administering the grants had to navigate through the local sponsored programs offices even though the grants were small. This was not anticipated when the project was first conceived. In some cases, sites started early and wanted to use funds retroactively. The mini-grant project provided challenges and benefits beyond the usual dissemination issues. But successful solutions of problems helped provide a solid foundation for the program.

4.1.1 Participation Data

Participation in PLTL through the WPA project included a large number of institutions, faculty, students, and workshop leaders. Some data are available for all of the 92 grants; but reports were not received from every grantee, so some data are based on fewer than the 92 grants. Since some grants were submitted with more than one principal investigator, more than 100 faculty participated. The total number of students involved in the PLTL WPA projects was greater than 14,000 and more than 1,000 peer leaders were included. Table 4.1 contains WPA grants by institution and discipline.

4.1.2 Selecting and Developing Workshop Materials

The development of appropriate materials is critically important to the success of Peer-Led Team Learning. The WPA grants provided an opportunity and incentive for participating faculty to consider carefully the kinds of materials to be used in the instructional process.

Table 4.1 Workshop Project Associate (WPA) grant data: numbers of institutions and disciplines

Basic data:	Number of grants	92
Type of institution:	Four-year colleges	42
	Research universities	19
	Two-year colleges	31
Discipline:	Chemistry (general, organic, biochemistry)	57
	Biology	20
	Mathematics	9
	Physics	3
	Other science areas	3
Workshop data:	Mean workshop size	8
	Mean workshop length	105 minutes
	Median workshop length	120 minutes

Materials for the PLTL workshops were adapted or written with a variety of needs and interests in mind. During the first four years of the project, several of the principal investigators wrote workshop materials for general and organic chemistry. These were published by Prentice Hall. PLTL adopters in these areas could use these in their workshops, as written or with modifications. Table 4.2 shows the sources of materials reported by the WPA grant recipients. (There is overlap between the second and fourth sources, with some respondents using PLTL among other sources.)

Table 4.2 WPA grants: workshop materials used

Materials used:	PLTL project materials only	9%
	Adapted PLTL project materials	33%
	Written entirely new	30%
	Written partly new, using several sources	29%

WPA reports cited the following about the development and use of workshop materials. These are direct quotes or very close paraphrases from reports. These comments show that the grant recipients looked carefully at the materials they had acquired or constructed.

- Materials were adapted to follow the sequence of topics for the course.
- Visual and kinesthetic activities were included, such as: ping pong balls; ionic compound puzzles; board games.
- Problems were arranged not linearly, but in ways that would allow more student self-determination.
- Materials were expanded to include test-taking exercises.
- More interdisciplinary and challenging activities were added.
- Materials were developed to emphasize conceptual understanding.
- Workshop materials were tailored to student and faculty interest.
- Workshop leaders assisted in adapting materials.

- Problems were set up to be multi-step and so that certain group activities could be carried out in the process of solving them.
- Cooperative learning activities were written to cover difficult concepts from both lecture and lab.
- Collaborative learning materials were adapted.

In summary, materials were customized in a variety of ways in order to: clarify concepts; increase the interest and motivational levels for students; accommodate professors' pedagogical priorities; foster cooperative work; and better meet the objectives of the workshops within the courses. This process of writing, editing, and re-working materials forced professors to consider closely: their goals for student learning; how students learn; how practice and cooperative learning helps students develop both skills and understanding; and how a leader can best facilitate progress as students work with difficult concepts and skills. The newly developed materials themselves were a contribution to the project, as implementers shared them with one another.

The PLTL founding group of faculty prepared a Guidebook and a Peer Leader-Handbook which was published by Prentice Hall and was very helpful in reducing the barriers to implementation as they were distributed free of charge to new and potential adopters of the model.

4.1.3 Leader Training

The selection and training of workshop leaders is viewed by most implementers as a central element in the success of PLTL. Ninety-one percent of sites reported holding initial sessions before the academic year to familiarize leaders with the method and their responsibilities. All sites reported holding weekly training meetings for the leaders. The time devoted to these weekly sessions varied from 60 minutes to two hours, with an average of 105 minutes.

Reports indicate that the following features of leader training were employed by most sites.

- The initial sessions were devoted to an explanation and discussion of the philosophy of PLTL, the role of the leader, and workshop dynamics.
- Topics covered in weekly meetings generally included: problems from the previous workshop; teaching/learning issues; preparation for the next workshop, including modeling of activities.
- Training was often undertaken in conjunction with a learning center or tutoring center of some kind.
- Some sites used "super leaders" to assist and sometimes direct the training sessions, and also to coordinate workshop logistics. The "super leaders" were experienced leaders who continued with the program, taking on added responsibilities after their first year.
- The leaders often worked through the entire workshop materials in the weekly meeting.

- In many cases leader meetings provided the basis for a one- or two-credit course. About half the sites reported that leaders maintained journals, completed teaching/learning projects, wrote papers, prepared posters, and in other ways analyzed and presented their experiences.
- Some geographical areas brought leaders together from several institutions for the initial training, and sometimes for the weekly leader meetings. Some sites accomplished this through electronic communications and distance seminars
- Sites varied in whether or not they followed the recommended procedure of not providing answer keys for workshop problems to the leaders.

The objectives and structure for the weekly meetings varied depending on the discipline, institution, and interests of the professor and learning specialist. In general the meetings were used: (1) to work through the material for the next workshop, reviewing both the conceptual background and the problem-solving strategies required; (2) to learn the theory and practice of teaching/learning strategies appropriate for that week; (3) to discuss individual problems encountered by the leaders and to anticipate issues likely to arise in view of the material, course timing, instructional priorities, and the like.

The earliest PLTL practitioners did not provide answers to the peer leaders, and as the project grew they encouraged new adopters to follow this practice. The conviction was that without answers the student leaders would better develop questioning strategies to use with the workshop group. It is recommended that part of the weekly training sessions be used to develop effective questions which the leaders use to guide students through the problems. This strategy alleviates most of the anxiety that peer leaders may feel in the absence of answer keys.

4.1.4 Collegial Interest and Dissemination Activities

Adopters of PLTL supported by WPA grants reported collegial interest in the workshops and said that they were themselves involved with dissemination. Table 4.3 summarizes findings in these areas.

The WPA reports describing interest among colleagues and dissemination activities reflected the project's breadth. In discussing dissemination, PLTL adopters considered strategies such as packaging and presenting information, the concerns

Table 4.3 Dissemination activities of WPA grantees

Interest among colleagues:	Little or none	13%
	Limited to the department	26%
	Within and beyond the department	61%
Dissemination activities:	Little or none	19%
	Campus level only	24%
	Regional meetings only	24%
	Regional and national meetings	32%

of potential adopters, existing and emerging networks, funding, and local priorities. The following are some WPA grant recipients' comments about dissemination:

- The website and project newsletter are helpful.
- There is no substitute for personal contacts and mentoring.
- Presentations at national conferences will keep PLTL in people's minds.
- Regional information and training conferences help promote adoptions.
- Training groups of experienced PLTL practitioners should visit interested institutions.
- Potential adopters should visit PLTL-using institutions.
- More quantitative data are needed showing the program's effectiveness.
- Materials should be available electronically for ease of use and adaptation.
- WPA grants of two or three years would help the program take root and yield more data.

4.2 Experiences of WPA Grant Recipients and Other Early Adopters

The WPA reports were concerned with local implementation and were limited in scope, in order not to be overly burdensome. Consequently, there are a number of program, dissemination, and institutional issues that were not covered in these reports. These issues include: how word of PLTL traveled; what characteristics of the project made it appealing to different groups and individuals; what previous experiences may have affected adoption of workshops; what obstacles had to be overcome to implement the program. These and many related questions are addressed using data obtained through site visits, phone interviews, and an online survey. We now consider PLTL's strengths and weaknesses as viewed by early adopters, and derived from these sources.

4.2.1 Strengths of Peer-Led Team Learning

PLTL adopters discussed the program's benefits for students, leaders, faculty, and the institution. The following are quotes or close paraphrases of the most commonly cited benefits.

Reported benefits to students:

- Workshops facilitate learning by encouraging student-student dialogue.
- The program engages students in chemistry in a way not before realized.
- Students: (1) feel more comfortable with chemistry; (2) become better problem solvers, thinkers, and aids to one another; (3) are less reserved and more willing to participate and to understand.

- Emphasis on problem solving gets students involved in how to think about chemistry problems, not just the answers.
- The method improves student confidence.
- Workshops help students find a study group.
- PLTL improves understanding of material and performance in the course. Students are more comfortable asking questions during workshops rather than lecture.
- Students who regularly attended the workshops performed above the class average.
- PLTL students, particularly leaders, developed maturity in analysis and communications that allowed them to do undergraduate scientific research and present their results at regional and national conferences. We believe this to be an important outcome. (Faculty at some institutions have argued against PLTL on the basis that the time leaders devoted to PLTL would prevent them from doing research. We find that this concern is not substantiated by the experiences of most PLTL-faculty as reported in interviews and site visits.)
- The program fosters cooperative learning, extra time-on-task, and extra time for discovery and kinesthetic lessons; it is pedagogically very sound.

Reported benefits to leaders:

- Leaders: (1) have an opportunity to experience teaching and decide to enter the profession; (2) enjoy an experience that can be the highlight of their college career; (3) become enthusiastic, courageous, willing to experiment, developing communications and problem solving skills.
- The program positively impacts our peer leaders as they assist in the acclimatization of freshman to the university.
- Peer leaders improved their own understanding of the course material, and some expressed a desire to become teachers.
- Helps leaders develop leadership skills and confidence.

Reported benefits to faculty and the institution:

- The ability to add an active learning component without changing all of the course structures makes the method easier to implement and disseminate.
- The program leads to a higher retention rate, and that is attractive to faculty and administrators.
- Along with the establishment of better relationships among the students, there is a higher level of confidence in both students and peer leaders.
- The program maintains an effective amalgamation of the principles of concept-building, metacognition, and cooperative generation of knowledge in a format that is both structured and flexible enough to be adapted.
- Higher grades and lower attrition are achieved. The program also bonds the students to the course and to their leaders.
- The program builds confidence in the leaders and provides a review of concepts and skills.

In summary, participating faculty list benefits in a number of important areas. They find that students are engaged in their learning, work cooperatively developing both conceptual and problem-solving skills, and perform better when tested. Leaders reinforce learning, gain new insights into the discipline, become experienced in facilitating learning, and gain confidence in their own abilities. An analysis of peer leaders' experiences at a private college can be found in an article by Tenney and Houck (2004). For their institution, the program helped address a number of priorities, particularly in developing a model for more effective instruction and developing a positive spirit about learning science.

4.2.2 *Implementation Issues*

When asked about weaknesses in the PLTL program, practitioners mentioned a wide variety of issues. Many of these were related to challenging circumstances in the implementation of workshops. A number of issues relate to use of the critical components of the PLTL model. Others involve the demands made on students, peer leaders, and professors. In order to explain the program more thoroughly, we list the following problems in italics and respond to each one. We categorize the issues in the following sections.

4.2.2.1 The Training, Supervision, and Role of the Peer Leader

- *Students do not like it that leaders are not tutors, and do not like not being given the answers or being told whether or not they are right.* The reasons for not supplying answers or telling students whether their answers are correct is to build understanding and confidence. Students should see clearly that when they have done their work carefully and correctly, their answers will be correct and that there are multiple ways of testing accuracy and developing confidence in their own answers At some sites, students worked through the problems without answers but were told when they had reached the solution or found the correct answer to a problem. In workshops that were observed, the leaders varied in their approach and skills.
- *Peer leaders, because of their busy schedules, were not able to meet often as a group.* This is a serious problem, occurring more often at commuter institutions. The program is seriously compromised if leaders do not meet regularly with the instructor to plan, review, and offer mutual support.
- *An important issue is the variability of the individual workshop experience. Perhaps this can be resolved by more extensive peer leader training or taping of workshops to view the strengths and weaknesses of individual peer leaders and groups.* Site visits revealed considerable variety in the skills and approaches of leaders. Even with training and supervision, there is variability in leader performance. This situation is in fact not very different from the variability among lecture experiences. Students find that professors differ greatly in the clarity of presentation, as well as the interest and motivation they generate. This is inevitably the case with workshop leaders as well. It reinforces the importance of careful selection, training, and supervision of the leaders.

- *Leader training and retention of leaders due to the institution only being a two-year school.* This is an on-going problem at community colleges. Some solved it by adding to their leader pool with students from neighboring four-year institutions. Others reported that many students took three or four years to graduate—due to outside employment and other commitments—and so there was in fact a pool of potential leaders available on site.

4.2.2.2 Materials: The Following Were Cited as Difficulties Related to the Availability of Appropriate Workshop Materials at Some Sites

- *Creating materials is time consuming.* This is true and is one of the reason why the first year of implementation is the most demanding. It is also why the PLTL leaders wrote and published materials for chemistry, and later for biology (Griswold et al., 2005).
- *A possible weakness might be the link between the materials and the lecture. If the materials are not closely related to the material the instructor covers, the students might not be motivated. Instructors must find time to closely coordinate the two.* The comment includes the solution. Many professors have said that having workshops imposed a certain rhythm on the course, in maintaining a desired pace and covering topics as planned. Workshop leaders often provide feedback to the professor, in written journals and verbally, about how students respond to the materials and workshop strategies.
- *Existing materials don't always fit and therefore, new materials must be developed which have not been field tested.* This is an issue. Materials are generally revised and improved over the first few years of use. But this in not very different from the use of a textbook or lecture material. Good instructors are always working to improve their instructional materials. The workshop leaders have been an important asset, providing feedback as materials are piloted and revised.

4.2.2.3 Funding and Organizational Arrangements Required for Effective Workshops

- *There is considerable demand on student manpower, institutional funding, and appropriate space.* This varies with the size of the course. It takes a lot of planning and scheduling to meet the needs of an introductory class with 200 students. But this simply reflects the cost of adopting real learning for each student. If an institution wants to save money by placing very large numbers of students in each class, it should be willing to spend a comparatively small additional amount to provide an environment that improves learning.
- *The biggest problem is that only one hour is available for PLTL.* This is a real problem. Workshops lasting only an hour are not the same as those lasting 90–120 minutes. Surveys revealed that students and student leaders perceived

significantly higher percentages of time spent on leader presentation and response in workshops of one hour than in those of two hours. Conversely, those in the longer workshops reported higher percentages of time spent on student problem solving.

- *No credit for the extra time invested by faculty*. Implementation of the program takes faculty time. The amount of time required diminishes after the first few years, but training and supervising leaders requires an ongoing time commitment. Colleges and universities should demonstrate that they place a high value on teaching and should reward faculty members who work to produce more effective learning. During a site visit, professors using PLTL in calculus courses at a large public east-coast university reported that the meetings in which leaders prepare for workshops have become very satisfying problem-solving seminars.
- *At this institution, education-related projects are not viewed as real research*. Expectations about research vary. Faculty members have to find out what will be considered "research," particularly if they are on a tenure track and not yet tenured.
- *One faculty member cannot sustain it alone. It must be tried in 'teams' - faculty working together*. Later in this report we will discuss the importance of implementation teams.

4.2.2.4 Student-Related Problems

- *There will always be difficulties dealing with unmotivated students*. This is true. The advantage of the workshops is that a lack of motivation, where it exists, is revealed, and can be addressed. Unmotivated students can easily be overlooked in lectures.
- *Some of the best students did not feel the workshops were necessary for them to be successful*. This issue has been addressed in a number of ways in terms of grouping, problem levels, and roles. One excellent student, a Rhodes scholar, in an advanced course said that the workshop problems were so difficult that no one could do them alone. Group work was required. Students in the course enjoyed the cooperative work and learned from one another. As this case demonstrates students will find the workshops more engaging if the problems are challenging. Instructors need to construct/select problems that are challenging for their own student populations. This is not a situation for which "one size fits all". In addition, successful performance on tests is not the only objective of education. Students should also acquire skills in teamwork, effective communication, and learning to speak the language of the scientific discipline. But these are not easily quantifiable.
- *Students expect workshops to replace study time outside the classroom and expect the workshop problems to be exactly like the exams*. Expectations should be clarified for each course component—lecture, study, workshop. Mazur (1997) has some excellent ideas on this. He asks students to write what they expect about different aspects of the course, and at a subsequent class he explains and clarifies each aspect of the course and what students should do as they participate in each area.

We agree that expectations should be clarified regarding all aspects of a PLTL course, including the role of lecture, workshops, study time, and textbooks.

- *Sometimes workshops break down into tutoring sessions despite the leaders' best efforts. A student falls behind and cannot contribute so the smooth discussion comes to a screeching halt.* Group dynamics and social issues arise continually and are addressed in the weekly sessions with leaders. In addition, it is extremely important that students keep up with course material and come to workshops prepared. If they do not, the learning and other workshop benefits will suffer.
- *Sports, other extra curricular activities and jobs interfere with the time commitment involved, especially since this was an evening program.... Scheduling of topics was difficult with multiple sections.... Evening exams occasionally conflicted with PLTL sessions.* Colleges do manage to schedule hundreds of courses, labs, activities, meetings, and events. It may be complicated, but can surely be done when there is a commitment to PLTL.
- *The method is self-limiting in terms of the number of disciplines that can employ it at any given location. There are only so many good students who could serve as peer leaders. The students who are qualified in one subject are probably qualified in several others as well. And so ultimately instructors will be pitted against one another in a competition for students to act as peer leaders in their discipline, as opposed to some other subject area.* This comment came from an institution implementing workshops in large classes of general chemistry and biology. While it is true that larger classes require more workshops and leaders, it is also true that there is a larger pool from which to draw. We have found that when workshops are successful, students tell the professors in other courses about them and that they would like to have them in other courses. In such cases, the issue of leaders tends to solve itself because as the workshops become more popular more students are experienced and ready to serve as leaders. But there are situations in which faculty and departments need to decide which courses would benefit by the introduction of PLTL. No one strategy can solve all problems and address all the needs of all students. Pedagogy needs to match the content.

It appears, based on the number and variety of issues listed, that there are challenges in implementing Peer-Led Team Learning. In WPA reports and interviews faculty have been unanimous in saying that implementing PLTL was demanding, requiring a significant amount of time and energy. There are many areas in which something can go wrong. The list of problems and issues reinforces the need for careful planning, often noted in these pages. But faculty members have also pointed out that it is very difficult to teach a course "that is not going well." Conscientious instructors sometimes have long lines of students to see them during office hours because in spite of their best efforts many students are not "getting it," in class. Those who have implemented PLTL successfully have found that the student gains amply compensate for the additional faculty work.

It is appropriate at this point in our discussion of teaching and learning to revisit an area discussed earlier, namely that the number of really good lecturers at a college is generally small—but no one sees this as a problem. When students fail to learn it is often considered to be their fault.

The above statement may appear to be gratuitous. But we have some evidence in the work of Lovitts (2001) who interviewed graduate faculty about doctoral programs. When asked about reasons for students departing from graduate study, faculty generally talked about a lack of motivation, preparedness, work ethic, loss of interest, and the like. When asked about factors leading to completion, in addition to student qualities, faculty members pointed to good advisement, quality courses, and faculty support. In other words they took credit for student success but did not want to share the blame for failure. We mention this issue to underscore the fact that academic support, built into a course, can be of great value to students and so faculty time is well spent as they resolve problems associated with PLTL.

4.2.3 *Impact of the WPA Program*

The Workshop Project Associate (WPA) grants were small, with a $5,000 grant and a matching amount from the institution. But faculty members wrote proposals, marshaled resources, and implemented workshops in response to the WPA opportunities. Investigators on the dissemination grant thought that the WPA grant would be a small incentive, perhaps able to push people in a direction they were already headed. But the grants accomplished more than that, Table 4.4 summarizes faculty responses about the role of the WPA grant.

In summary, 68 percent of the respondents said they would not have implemented the workshop program without the $5,000 grant. The report form also asked about plans to continue PLTL with the following choices: (a) expand the program; (b) continue the program as is; (c) review the program to make a decision about the future. Responses are summarized in Table 4.5.

Most institutions that used a WPA grant to implement PLTL planned to continue or expand the use of workshops. After a few more years, some of the adopting institutions had serious difficulties and some abandoned the program. Case studies including site visits, interviews, and further analysis were used to investigate these events. They will be discussed in subsequent chapters on institutionalization.

Table 4.4 Responses to the item: would PLTL have been implemented without the WPA grant?

Discipline	Yes	No	Not sure
Chemistry	17	33	3
Biology	2	19	2
Math/Physics	1	3	1
Total	20	55	6

Table 4.5 PLTL continuation plans

Discipline	Expand	Continue	Review
Chemistry	18	35	0
Biology	5	16	1
Math/Physics	1	5	
Total	24	56	1

Based on the reports summarizing the experiences of WPA adopters, we can say that the mini-grant approach was an effective method for encouraging individuals to adopt the workshop method at least on a trial basis. Requiring matching funds guaranteed a level of institutional support. As the data indicate, most of those piloting PLTL planned to continue the workshops. The pilot grants also made the project visible to other faculty members who often became interested, or at least curious about the program.

The entire mini-grant effort seems to be a good example of a diffusion or dissemination effort taking advantage of the existing dynamics. Dearing (2004) finds that there have been few studies looking at these natural processes. In the case of PLTL, the natural process in which a professor notices, learns about, and then adopts a practice was amplified by the use of the WPA grants for implementation. Professors and institutions were given the basic resources to initiate something that was attractive to them and that, based on their reports, they would not have undertaken without the grant. Dearing, comparing several pedagogies, states, "PLTL seems more sophisticated than the other pedagogies in its means of dissemination. It involves multiple stakeholders all who are connected with faculty as potential levers for tipping faculty behavior in a new direction" (p. 93).

4.3 Summary

The WPA mini-grants were successful in disseminating PLTL for the following reasons: (a) funding became available for leaders and the development of materials; (b) grant writing and institutional matching required and insured administrative support; (c) awarding of the grant attached the participating institution to the PLTL national network; (d) the 92 WPA participants provided the project with a wealth of implementation and dissemination data; (e) many of the WPA faculty became the next generation of leaders in adapting and disseminating the method.

The WPA study demonstrates that institutions can use small grants at their own institutions to try new strategies in teaching/learning and bring about major changes in the classroom with relatively small funds. A table appended to this report, lists the institutions obtaining WPA grants.

Chapter 5
Indicators of Institutionalization

Rogers (2003) defines sustainability or institutionalization as, "the degree to which an innovation continues to be used over time after a diffusion program ends" (p. 183). With some changes in vocabulary, this definition applies to the PLTL project. In most cases initial implementation was made possible, at least in part, by external funding. But the institutionalization of PLTL depends on a variety of factors of which funding is just one.

Scheirer (2005), identifies five important factors influencing sustainability: (1) program flexibility, (2) a champion, (3) fit with the organization's mission, (4) perceived benefits, (5) support from stakeholder organizations. All of these have already been noted as significant factors in our discussions of Peer-Led Team Learning, and they will be further discussed in this chapter. Stakeholders for PLTL are faculty, students, student leaders, and administrators—and to a lesser extent parents, alumni, and the interested public. We will consider issues of sustainability in several other educational initiatives to see what happens when one or another factor is particularly strong or weak. We will also propose our own set of critical components essential for the institutionalization of an educational initiative.

5.1 Experiences of Other Initiatives

Only time can really tell whether a new program, curriculum change, or method of teaching will in fact persist. Many educational initiatives, even those with proven effectiveness, do not endure and it is useful to consider several programs in order to gain insights into the issues, obstacles, and supports that surround an initiative.

5.1.1 The Keller Plan

In 1968 Fred Keller a psychology professor at Columbia University published his Personalized System of Instruction (PSI). Keller believed that for most students,

L. Gafney, P. Varma-Nelson, *Peer-Led Team Learning: Evaluation, Dissemination, and Institutionalization of a College Level Initiative*,

lectures were not the best format for learning. As described by Cracolice & Roth (1996), the Keller plan used self-pacing, allowing students to learn at their own rate, with individualized testing, peer proctors, few lectures, and texts for independent study.

In the Keller Plan, students receive a course outline and materials, they study, and take a test. If test results indicate that they have not mastered the material, they receive tutoring—individually or in study groups. They then retake a new version of the test and, after mastery is demonstrated, move on. Although first introduced in psychology the method became widely used in chemistry.

The approach is demanding in time and energy for both instructors and students. But the success rate, and the generally positive atmosphere of the program was such that both instructors and students reported more favorable attitudes than in the case of traditional courses.

In spite of its success, a 1986 survey found that only 23 of 43 adopters were still using the Keller Plan, and adoption by new faculty was rare. Why did this effective method fall into disuse? Cracolice points to several factors. First, the approach is labor intensive, requiring carefully developed materials, recruitment and supervision of peer proctors, and individualized testing. Second, some instructors balk at exchanging the role of lecturer for that of instructional manager, believing that the professor's role in clarifying concepts, making connections, and modeling problem solving is very important for student learning. Third, some students procrastinate in this format and do not maintain a suitable pace. Although these issues can be addressed, the method has generally been abandoned.

In summary, we can say that the Keller plan did not demonstrate sufficient flexibility and its benefits were not great enough to sustain it in the face of serious challenges. In addition, although the plan had apparent advantages for learners, these did not translate to clear benefits regarding institutional priorities.

5.1.2 Calculus Reform

In the mid 1980s, a variety of issues confronted college calculus instructors and those teaching in client disciplines. As listed by Tucker & Leitzel (1995), issues included: too few students successfully completing calculus; students mindlessly working through algorithms; under-prepared students entering calculus classes; and mathematics lagging behind other disciplines in the use of technology. With support from the National Science Foundation and others, concerned mathematicians and educators developed new approaches to calculus instruction and learning.

What emerged was a radically new curriculum, characterized by the integration of intuitive, formal, and applied mathematics. As far as possible, each topic was treated from several perspectives—numerical, verbal, graphic, and algebraic. Technology was used more extensively, both to reduce tiresome computation, and to introduce real-world applications, not only from physics, the traditional applications area for calculus, but also from biology and the social sciences. Textbooks and other materials were prepared and professional development workshops were conducted across the country to familiarize faculty with the new methods and benefits to be derived from their use.

Due to extensive dissemination efforts and the natural attractiveness of the new approaches, reform calculus methods spread quickly. The success of the approach is attributable to several factors. Most importantly, the method has had proven success. Studies show that students not only gain a better grasp of the conceptual and applications dimensions of calculus but even outperform traditional groups in algorithmic areas (Tucker & Leitzel, 1995).

But the success of the calculus reform may have been possible because many traditional educational dynamics did not change. The lecture, use of textbooks, homework, test taking, and the role of the instructor all remained intact. The material itself and to some extent the manner of learning changed but this was accommodated within the time-honored framework. Again, to summarize in view of the factors for institutionalization listed by Scheirer, reform calculus had champions, nationally and at most adopting institutions. It could be implemented with considerable flexibility. There were perceived benefits in terms of student interest and greater satisfaction among client disciplines—important stakeholders.

5.1.3 PLTL Issues

As an educational initiative Peer-Led Team Learning lies somewhere between the Keller plan for personalized instruction and reform calculus. The introduction of workshops as practiced in the PLTL model requires changes in organizational structures such as scheduling which is no small matter, including as it does the factors of time, space, group size, faculty assignments, and coordination with other classes and activities. The method also requires new personnel and funding with the introduction of peer leaders. The method also requires some changes in the role and responsibilities of the lecturer. Finally, adopting the method calls for training in a new pedagogy, and this may be a new experience for faculty.

These changes are more difficult to introduce and to sustain than changes in instructional materials and teaching strategies that involve just the professor. We might say that reform calculus required faculty to take a new look at the discipline, but Peer-Led Team Learning required them to take a new look at themselves and their professional roles. The demands for self evaluation and pedagogical transformation are not as dramatic as in the Keller plan because lectures are not abandoned. But offering an undergraduate, as peer leader, a formal role in the course structure and in the teaching/learning process is a significant change in faculty behavior.

5.1.4 The Professor's Role

It will be useful to consider the role and self-image of the college professor. Preparation for the professoriate is arduous. It usually takes five to seven years to earn the Ph.D. degree, and this is often followed by a post-doctoral experience for those seeking positions at more prestigious institutions. The prize, in terms of an academic position, pays reasonably well, and offers a high degree of independence in how one works within the established framework—of teaching, research, and service. The position also includes a high level of security once tenure is achieved.

The expectations and requirements for teaching are clear—although they vary by institution. The more prestigious the institution, the less teaching time is required or expected because research is a priority. Grants can bring funding that may be traded for further reductions in teaching, in order to devote time to research. Faculty engaged in research generally gain their greatest satisfaction and recognition from the findings, publications and other rewards resulting from this activity.

In general, teaching at the college level means lecturing, providing office hours for individual student questions, administering tests, evaluating student work, and assigning grades. For large classes, professors generally have teaching assistants to help with grading, and sometimes to hold recitation classes in which students can ask questions. At well-funded research universities, professors may teach one or two courses a semester; at community colleges it may be five.

Within this framework, faculty have a great deal of freedom in how they teach, and they generally exercise this freedom within the lecture format. Many professors are justifiably proud of their skills in presenting concepts clearly, explaining problem-solving techniques, and at times inspiring students. In recent years, many have adopted the use of PowerPoint, and other technologies as well as demonstrations, and clickers, activities in which students interact with one another during the lecture. But the professor remains in charge and at the center of the process.

A number of very important questions therefore arise as faculty members consider the continuation of a PLTL program after initial implementation. These include the following.

- Are the benefits to students sufficient to justify the efforts and expense required to continue the program?
- Do the benefits to faculty outweigh the cost and possibly reduced attention to other professional areas, such as research?
- Do faculty members accept modifications of their own roles that include sharing responsibilities with peer leaders?

The role and self image of the college professor are very interesting and important to PLTL and are discussed at several points in this study.

5.2 Administrators' Views About Institutionalization

Since PLTL requires cooperation and support beyond the individual course, Gafney considered the experiences and outlook of administrators to be important in analyzing sustainability issues. After PLTL had been in use for about five years, we conducted phone interviews with 12 administrators from 10 adopting institutions, asking about their views of the project and its sustainability at their institutions. We believed that successful implementation and institutionalization depended on the support of department heads, deans, vice-presidents and provosts who saw the program as enhancing learning and furthering the goals of the institutions. Interviews with administrators uncovered the following.

1. A fit with local priorities and previous experiences can provide the opportunity for a good start. Administrators were asked whether institutional priorities provided an environment suitable for Peer-Led Team Learning. In response, almost all of those interviewed reported involvement in teaching/learning initiatives and curriculum revision that preceded and in some cases previewed the adoption of PLTL.

   - At Portland State, the Freshman University Studies Program grouped 30–40 students with a peer mentor. These mentors met regularly with students; worked with them on computer, library and other skill areas. The mentors were generally seniors and were paid. The administrator whom we interviewed had participated in this program, and its success in using peer leaders disposed him to recognize benefits in PLTL and to support the program.
   - Faculty from the University of Portland, Coastal Carolina, and Goucher College had been experimenting with new pedagogies. Professors had taken sabbaticals to investigate new programs. The knowledge, energy, and innovative spirit of these faculty members impressed administrators who were then ready to endorse PLTL.
   - At Indiana University-Purdue University Indianapolis (IUPUI), a forum was held with administrators discussing gateway courses and retention: 62 percent of freshmen were the first in their families to attend college and retention rates were low. The general chemistry instructor was conscientious about improving his teaching, and even held student focus groups to discuss lectures. These institutional and professional priorities led to the adoption of PLTL, with marked improvements in retention in the course.
   - Evidence of improved student performance and retention was a priority for deans, provosts, and presidents. They wanted to see academic success. But many new programs, even those that are genuinely beneficial, find it difficult to produce convincing data. Innovative curricula sometimes introduce new goals and this makes it difficult to compare the outcomes with those of programs that were replaced. PLTL, however, has found that student performance can and does improve as measured by traditional class tests, whatever these might include, and many sites collected data supporting this goal. Administrators generally found these data persuasive, but they often wanted new studies on their own campuses.
   - Deans and presidents were often invited to special PLTL activities, demonstration workshops, poster sessions and the like. In interviews they indicated that they were impressed with the poise of student leaders whom they saw as potential ambassadors in recruiting for their institutions.

2. The introduction of PLTL at an institution requires administrative support in rewarding peer leaders, often, but not always, through funding in the range of

$500 per semester. In terms of a university's budget this may be a small amount, but funding is often an issue. The administrators interviewed recounted a variety of successful strategies for funding PLTL.

- Existing funds such as institutional support for tutors, learning centers, work-study and student service activities were sometimes directed toward PLTL, as long as students satisfied the criteria for the program funding.
- State funds for designated curriculum initiatives, as in California, were used for PLTL. These funds were expected to recur and to provide a reasonably secure base for the program.
- Financial support for PLTL was sometimes included in other grants such as the NSF-sponsored Louis Stokes Alliance for Minority Participation (LSAMP), and the NSF College Curriculum and Laboratory Initiative (CCLI); NIH-sponsored programs for student research; and grants under the Fund for the Improvement of Post-secondary Education (FIPSE). These programs have had primary objectives related to minority students, women in science, active learning, research, teacher preparation, or some other area. But they have included PLTL as integral to the attainment of the project goals.
- Deans and presidents often had access to discretionary funds that they directed toward PLTL.
- The best and most secure funding for PLTL, or any program, was of course inclusion as an item in the departmental budget.

Some institutions experimented with approaches to support peer leaders without funding, including the following.

- Students were not paid the first time they served as leaders, but were paid as leaders for subsequent semesters.
- Peer leaders received credit for attending the weekly workshop preparation sessions and for leading workshops. This was different from arrangements in which students were paid for workshop activities and received credit for a separate pedagogy course.

3. Local dissemination and institutionalization often complement one another. The same activities that contribute to successful dissemination can also move a project toward institutionalization. Administrators noted the following factors at sites where PLTL has been successfully introduced and gained momentum.

   - PLTL information flows informally among faculty. This flow seems to be more effective when: faculties are smaller; there is physical proximity among offices; working connections already exist among faculty members and departments; and there is a serious interest in the study of teaching and learning.
   - Successful workshops produce satisfied students who recognize what PLTL has done for them, and talk about it. Faculty members listen to them. Administrators report that they became aware of PLTL's successes through both formal and informal communications networks.
   - Following initial workshop success, faculty members present the method, their experiences, and findings at local meetings, on campus or at regional

conferences. Administrators are influenced and involved because they want to see improved student performance and faculty interest in pedagogy. Recognizing these benefits, they look for secure funding.

So dissemination across a site tends to: get people involved; address pedagogical issues; create ideas and sources for compensation; make PLTL part of the way things are done. These are important steps toward institutionalization.

4. PLTL can have a bridge-building and bonding effect. Administrators repeatedly stressed an area, that had not frequently been mentioned by faculty, as an important benefit of PLTL, namely the various ways in which the program created new connections and networks.
   - Community colleges and urban commuter school administrators reported that PLTL helped bring students together for academic work in ways that carried over to other courses. Students began to see the institution as a location for informal learning and study groups, not only for attending lectures.
   - There was general agreement that workshop leadership provides an excellent experience for students with an active or potential interest in teaching. Workshops fit very well with contemporary approaches to teaching and learning, and can motivate students toward teaching careers at the college or school level.
   - When a faculty is interested in discussing teaching and learning, PLTL has provided a focus and an area of program agreement across departments, and has been a fertile area for professional development.
5. The final question of each phone interview asked administrators what they would like to see in a proposal to fund a PLTL program. The following are some of the more important elements suggested:
   - A clear description of the program, its distinguishing characteristics, and how it differs from other programs
   - Evidence of improved student performance
   - A link to retention and recruiting
   - Description of a structured program to guide peer leaders
   - A commitment from faculty to direct, manage, and evaluate the program
   - A description of the benefits to peer leaders
   - Evidence of collegiality and bridge building, if possible

5.3 Indicators of Institutionalization

Based on interviews with administrators described above, on the experiences of PLTL adopters, and the analysis of many contacts with faculty, we developed the following list of indicators to determine PLTL's sustainability in a given setting or institution.

1. *Fidelity to the model.* The PLTL model comprising six critical components has been mentioned throughout this study. Its clarity made dissemination possible,

and we have repeatedly found that when one or more of these elements is missing or not adequately employed, the workshop program will be compromised and vulnerable when threatened. We conclude, then, that fidelity to the model is essential for implementation and is at least a starting point for sustainability.

2. *Funding and administrative support*. Some types of funding are more permanent than others. PLTL is more secure when it has a regular budget line, or is part of the budget that is routinely renewed, such as departmental funds for tutors or undergraduate teaching assistants. Similarly, acknowledgement and support from the institution's administrators lead to a greater likelihood that PLTL will persist. The program gains security and participating faculty feel that their efforts are appreciated. This translates to greater confidence among all involved.
3. *Perceived success*. We have noted a number of times in this report that given the amount of work involved in the implementation of PLTL, faculty and administrators want to demonstrate success at their institution. This is why so many comparative and norm-based studies have been and continue to be conducted. With changes in teaching assignments and in faculties, it is critical that each person taking responsibility for a PLTL course recognize the benefits and be committed to implementation of the complete model.
4. *Fit with the institution's mission and practice*. All colleges and universities say and believe that they are strongly committed to high quality teaching and learning. But some institutions act on this conviction more than others. When this commitment is exhibited in hiring, spending, pedagogical research, and program development, PLTL is more likely to flourish. When an institution and its faculty promote student initiative, independent thinking, and cooperative approaches, PLTL finds a good fit.
5. *A core group of committed faculty*. Like all programs, PLTL ultimately comes down to people. If the faculty members implementing the workshops are absolutely convinced that the program is beneficial, that it makes a significant difference in learning, then they will work through any problems and will convince colleagues of the program's value. We have seen that a group or team is needed because a program depending on one person is extremely vulnerable.

These five factors are strong indicators of successful implementation and of sustainability. We now look at case studies to explore sustainability in practice. At the end of the chapter we will return to this model for institutionalization.

5.4 Case Studies: Sustainability Issues

Practitioners may explain and promote a particular model, as has been the case with PLTL, but those who listen and adopt the method bring their own institutional cultures and priorities to the implementation. The following cases demonstrate what has happened at different sites implementing PLTL.

5.4.1 Implemented by a Team

Implementation by a team of faculty members, from one or several institutions, has usually been the result of previous departmental efforts to improve teaching, learning, and the curriculum. When a group has attended PLTL professional development workshops, they spent the last half day, of a 2.5 day conference, in a team planning session. They formulated action plans for the training of leaders, preparation of materials, and scheduling. In the process of implementation the team members were then able to confer, delegate tasks, solve problems, and provide a united front in explaining the method to other faculty and administrators. This collaborative approach provides a more positive outlook for sustainability.

- *Urban cooperation.* Faculty members from three urban institutions—a state university, a private college, and a community college—had cooperated on a previous grant and were attracted to PLTL. All three institutions received WPA grants. The private college is a small, private, four-year institution and the collegial approach to implementation was particularly noticeable there. The chemistry and biology departments worked together with strong support from the dean, and assistance from a learning specialist. They monitored results, found marked improvements in student performance, published their findings and as a result the program seemed to be virtually institutionalized. But collegial activity at the college level, particularly across departments, generally depends on individuals, not on permanent structures. When one of the original implementers of PLTL at the private college retired, the successor—as of this writing—was using a very modified form of PLTL.
- *State university.* At a state university, enrollments had been increasing and grades in general chemistry had been dropping. The chemistry department was looking for ways to improve student performance. PLTL was adopted and implemented as a team activity by those teaching general and organic chemistry. The department chair used statistical measures to demonstrate improvements in student performance. Other faculty members took different implementation responsibilities thus sharing the burden involved in: recruiting leaders, organizing and scheduling; preparing materials; developing pedagogical strategies. This teamwork and informal delegation of tasks was a major factor in furthering the program. Most recent data indicate an ABC success rate of 82 percent for PLTL students contrasted with 58 percent for those not in the program. With solid departmental funding, PLTL appears to be well on the way to institutionalization at this site.
- *Offshore university.* Chemistry faculty at an American offshore university independently developed a tutor-mentor program with most of the features of PLTL with the added features that the peer leaders were expected to attend classes of their workshop students and were trained in mentoring as well as tutoring techniques. After implementing the project for a year the faculty became acquainted with the PLTL project and adopted the method. The program achieved considerable success with an ABC success rate of 69 percent (n = 424) for participants and 53.5 percent (n = 1,425) for non-participants, over seven semesters (Báez-Galib et al., 2005).

5.4.2 Implementation by a Single Individual

In many cases PLTL was implemented by an individual who was looking for more effective teaching and learning strategies. But the lack of collegial support and practical cooperation has often made it difficult for a single individual to sustain the initiative. In this case the program does not gain the advantage of review from different perspectives. The burden on one person is great. Finally, an individual can succeed with departmental support and funding, but when that individual leaves the institution it is unlikely that PLTL will continue.

- *State university*. A state university was partner in one of the NSF Adopt and Adapt grants. A faculty member there became thoroughly committed to the method, introduced it in his sections of general chemistry, and became active in the national PLTL network. Students and peer leaders responded favorably, and data indicated the success of the method particularly with under-prepared students. But other faculty in the department, while admitting the need to engage students more effectively, said that they wanted more evidence regarding: improved student performance at the university and other sites; the value of the program for the leaders; and the time that implementation would take from research. As a result there was no expansion at the university.
- *Urban community college*. PLTL was introduced in certain biology, chemistry, and mathematics courses at an urban community college with the support of a WPA grant and through a grant-sponsored consortium related to teacher preparation. But implementation varied and the three departments worked in isolation from one anther, with varying levels of fidelity to the PLTL model. The problems associated with selecting and training of leaders, scheduling workshops, developing materials, and securing funding became burdensome for one of the professors who was also working toward tenure.
- *Private college*. PLTL was introduced by a member of PLTL's founding group at a private college in the general and general organic, biological chemistry (GOB) course for nurses. The response of students and leaders was highly positive, and scores improved, particularly on the ACS exams. Within a year, PLTL was also implemented by two other professors, in the general chemistry and organic courses. After the PI left the institution, PLTL was retained in two of the three start-up courses.
- *Implementation facilitated by the department chair*. The implementation of PLTL can be particularly effective when initiated and directed by the chair. We have records of two instances in which this worked well at state universities. One in biology (Gaines, 2001) the other in chemistry. In each case, the chair used persuasion, funding, and organizational expertise to plan and implement the program in several courses. The chairs also made use of the PLTL network to plan and implement effectively.

5.4.3 Support and Direction of Workshop Leaders

The selection and support of leaders is perhaps the most important of the critical components for PLTL. The leaders must receive support and direction so that they develop competence and confidence and see themselves as part of the course and the teaching/learning initiative. The City College of New York and the University of Rochester were among the earliest implementers of peer-led workshops. At both institutions learning specialists participated in the program, and considerable attention was paid to the selection, training, and supervision of peer leaders.

The learning specialists at these two institutions, working with one another and with the participating faculty, developed a course for the leaders covering instructional issues, learning theory, and activities in group dynamics. They went on to publish a handbook for peer leaders (Roth et al., 2001), and contributed to the guidebook for faculty (Gosser et al., 2001).

Although the program recommends participation by a learning specialist to assist with pedagogical issues, at many institutions such a person is not on staff or is not available. PLTL has been implemented successfully at these institutions, without the aid of a learning specialist. The faculty members implementing the program at these institutions with and without learning specialists placed a high value on the peer leaders, not only as facilitators of student learning but also as partners in the workshop enterprise. In this capacity, the peer leaders brought ideas about teaching and learning to the project and also developed a strong camaraderie in their work with faculty. As leaders nominated candidates to be future leaders, bonds were forged among the leadership group. Loyalty to the program was further enhanced as leaders from these institutions often traveled to conferences and presented the method to potential adopters. These students became some of the most important figures in disseminating PLTL.

5.4.4 Issues in the Direction of Leaders

Although PLTL adopters understood the importance of the selection, training, and ongoing support of the peer leaders, local circumstances sometimes interfered with thorough implementation. In these cases, the leaders sometimes noted a disconnect between the lecture and workshop, or they failed to acquire the range of skills needed for successful implementation.

- *State university*. At a state university, PLTL was implemented by two members of the chemistry department obtaining a WPA grant. The workshops were adopted as one of several course features that could be selected by students with a grade-weighting system permitting them to customize the course, within certain parameters. After several years, one of the original implementers left, the other became chair with many added responsibilities, and several new faculty were hired. The result of these changes was that less time and expertise was available for weekly meetings with the leaders. Added to this was the fact that a large number of students were commuters who added employment to their college responsibilities and so it was difficult to schedule weekly meetings for the leaders. Finally, the highly flexible style in which PLTL was implemented did not provide a solid foundation for the program. In view of these factors, the workshops did not appear to have strong positive results and consequently were abandoned in chemistry courses.

- *Private university*. At a private university, a number of faculty members from different departments adopted PLTL. But they did not seem to work together and so a number of leader-related issues arose. The workshops were generally drop-in sessions with attendance said to vary from 3 to 30. The peer leaders behaved like recitation leaders—fielding questions, demonstrating problem solving techniques, and helping students prepare for tests—rather than facilitating small group learning. Chemistry faculty said that they had difficulty attracting leaders because of other demands on student time, such as research. Physics faculty said that only about half the leaders usually attended the weekly meetings. The leader training sessions were handled as a for-credit course one year but some faculty thought this was counter productive, with some able students declining to become leaders. A number of these difficulties may relate to the drop-in approach and the fact that leaders are not highly valued by the program. Finally, a non-tenure track faculty member initiated and supervised PLTL.

5.4.5 Mandatory Versus Voluntary Workshops

Although the PLTL model and dissemination activities propose a model with workshops mandatory for all students in a particular course or lecture section, data collected through surveys, site visits, and reports indicate that only about 50 percent of the adopters do in fact require workshops for all students. Although it has been predicted that participation based on student choice would undermine the effectiveness of the workshops, this has not necessarily been the case in practice (Wamser, 2006).

By "student choice" we mean that students elect to participate in workshops for the length of the course, and after making the decision they are required to attend. There are two main reasons given for participation based on student choice: first, limited resources in terms of funding, availability of leaders, and organizational arrangements; second, the belief or experience that uninterested students will make the workshops less effective for others. As noted above, the program tends to become adapted beyond recognition and is not successful if workshops are held as drop-in sessions.

5.5 Online Survey on Implementation, Dissemination, and Institutionalization

During the 2004–05 academic year, PLTL implementers were asked to complete an online survey. We were interested in issues of implementation, dissemination, and institutionalization. After three notices, 60 completed surveys were obtained from faculty at 40 institutions. The types of institutions are shown in Table 5.1.

The information received through this survey overlapped with what had been learned from the WPA reports, but there was also new information. This survey tested the five hypotheses for institutionalization, listed earlier in the chapter: (1) Fidelity to the model: (2) Funding and administrative support; (3) Perceived success; (4) Fit with the institution's mission and priorities; (4) A core group of committed faculty.

5.5.1 Comparisons Between Lower and Higher Ratings on Institutionalization

Table 5.2 shows the means on a 1–10 scale for responses to seven survey items. The first column lists items from the survey on institutionalization. The second column lists the mean for that item for respondents believing that PLTL is less likely to endure at their institutions. The third column gives means for respondents believing that PLTL is more likely to endure at that institution. The fourth column gives the differences between the means of "more likely" and "less likely" respondents.

The table throws light on the hypotheses for institutionalization. The first two items are about initial implementation and the use of the model. There is little difference in response to this item between the means of those who are more confident about institutionalization ($r > 5.5$) and those who are less confident ($r < 5.5$). These results, along with the fact that PLTL implemented according to the model was sometimes abandoned, caused us to question our hypothesis that implementation according to the critical components, by one thoroughly familiar with the model and committed to its success would of itself be a predictor of sustainability. This

Table 5.1 Institutionalization survey: respondents by type of institution

Type of institution	Percent and number responding
Community College	15% (9)
Four-year private college	18.3% (11)
Four-year state college	21.7% (13)
Private research university	18.3% (11)
State research university	26.7% (16)

Table 5.2 Mean ratings for selected survey items

	Likelihood of institutionalization r = rating		
The likelihood that PLTL will be used at our institution in 10 years on a scale of 1–10 (1 = not at all likely; 10 = definitely likely) is: Survey Items	$r < 5.5$ (n = 18) Less likely Mean	$r > 5.5$ (n = 42) More likely Mean	Diff
Fidelity to the model			
1 A. Peer-Led Team Learning was first adopted at this institution by someone thoroughly familiar with the program and strongly committed to its success	7.8	7.6	−0.2
1 B. Implementation of PLTL followed careful use of the six critical components	7.8	7.9	0.1
Funding and administrative support			
2. Administrative support was forthcoming for PLTL, including funding and access to appropriate time and space	6.6	7.8	1.2
Perceived success			
3. PLTL is viewed as successful by administrators and faculty familiar with the program	6.6	8.4	1.8
Fit with the institution's mission and practice			
4. There has been a good fit between PLTL and other departmental and institutional priorities	6.4	7.9	1.5
A core group of committed faculty			
5 A. After initial implementation, PLTL was adopted by other faculty	3.8	7.2	3.4
5 B. PLTL faculty have participated in local and/or national dissemination activities	7.1	8.6	1.5

does not seem to be the case. As we have seen PLTL can be carefully implemented, but unless other important elements are also in place, the initiative may not last.

Understanding the model and using it properly are important for initial success. But respondents to the survey who predicted sustainability and those who did not gave virtually the same ratings to the level of implementation and quality of the initial implementer. Fidelity to the PLTL model is required for successful implementation and is therefore a necessary but not a sufficient condition for institutionalization. As a test for institutionalization, implementation according to the critical components might be said to have few false negatives, but might easily have false positives.

The next items deal with the program in the context of the administration, perceived success, and institutional priorities. For these there are significant differences between the mean ratings of those who predict institutionalization and those who do not. We thus found strong support for these hypotheses for sustainability.

The differences between the means for Items 3, 4 and 5A are statistically significant ($p < 0.01$). The greatest difference in means is in response to the item, "After initial implementation, PLTL was adopted by other faculty" ($p < 0.0003$). These results are very interesting. They say in effect that local diffusion is a key factor in institutionalization. When the program grows, it is more likely to endure.

5.5.2 Fidelity to the Model and Sustainability

In 2000 we issued a report indicating how a number of sites had difficulties in implementation that could be traced directly to weaknesses in using the critical components—for example, in the kinds of materials used, the size of the group, the training of leaders, the involvement of the professor. At that time, one institution was cited as following the model closely and achieving noticeable initial success. In 2005 one of the original implementers from this institution retired and was dismayed by what was happening. The situation is described in an email as follows.

> *I am retiring May 31st. There is one in my department doing what he calls PLTL but it is more like discussion sections with student leaders answering the questions. This is also my opinion of what another department is doing. They do not train their leaders nor meet with them weekly to discuss their groups. I am reminded of the article you wrote listing six essential items for successful implementation. I think you nailed it. Many seem to want to use the PLTL moniker without doing the work involved that those six items entail. In another department, the professor was on sabbatical this past year and for the first time in the evaluations of PLTL the students were not as enthusiastic about it. The professor in charge of the workshops in that department was not the one teaching all of the sections of lecture so that the exams did not reflect workshops.*

The history of PLTL at this institution, recounted above, reveals some of the details about why successful implementation according to the model is an ongoing issue. Each new faculty member or department working to implement the model, should be made aware of the critical components and the importance in seeing that they are in place. Faculty turnover is a huge issue, not unique to PLTL. When an outstanding lecturer or lab supervisor leaves an institution, there is a void until someone emerges to take his or her place. The difference is that in the case of PLTL considerable energy may be needed to revive or restart the damaged program.

5.5.3 Institutional Priorities and Support as Indicators

We noted from the outset of PLTL that some of those who became interested in the model had been searching—as individuals, or in small groups, or as ambassadors for a department—for teaching/learning strategies that would engage and motivate students, and lead to more effective learning. Survey items 3 and 4 in Table 5.2 relate to these issues, and those who have higher expectations of institutionalizing the program gave significantly higher ratings to the fit of the program with local priorities and with institutional support.

A Midwestern university, mentioned in the case studies, provides a particularly good example of this. The chemistry department planned together and stayed together in implementing PLTL. Permanent departmental funding was made available to support the program, in particular by providing stipends for the leaders.

5.5.4 Perceived Success and Institutionalization

Table 5.3 shows faculty comments about benefits of PLTL, again sorted according the respondent's rating of the likelihood of sustainability for 10 years. The left-hand column shows individual ratings on a scale of 1–10 (most likely) to the item, "It is likely that PLTL will be at this institution in 10 years." These are selected from the survey that was the basis for Tables 5.1 and 5.2. In this we selected from

Table 5.3 Institutionalization rating and major benefits of PLTL

	The column on the left gives the individual respondent's rating (1–10) regarding the likelihood of sustaining PLTL for 10 years. The column below provides individual comments about the benefits of PLTL
	PLTL benefits from those giving low ratings for institutionalization
2	Growth of leaders. Helping students at most levels that invest time in the program. Increasing student motivation
	We had short-term success with PLTL until the funding was cut
	Student satisfaction with the course
3	To student leaders. To a handful of students in the class who really appreciate and enjoy the small group setting
	Student leaders really did enjoy the connections between themselves and students in the workshop settings. Leaders developed a real confidence and did recommend continuation
	Improved student performance
4	(1) A great source for tutors and team leaders. (2) More lower ability students learn better. (3) Students feel more confident about what they have learned and retaining it for future reference. (4) Students saw this as a way to enhance their study time
	Boost in student morale
	PLTL benefits from those giving the highest ratings to institutionalization
9	Student performance in a traditionally difficult class sequence
10	Student grades and student engagement with the course material
	Improved student problem solving skills, better understanding of concepts
	Improved student learning
	Increased student learning and improved student attitude towards the courses
	Improvements in students' understanding of material. Development of mentors
	Students working on group learning, developing problem solving skills, and better grasp of difficult concepts
	For the students: Improved content mastery Higher grades. Improved team and people skills
	For the faculty: Better job satisfaction, based on knowing that PLTL improved conceptual understanding of the courses.
	For leaders: Deeper conceptual understanding, which was helpful in some courses they are taking, development of team, leadership and conflict resolution skills, better study and time management skills
	For the learning specialist: Being part of an exciting and new way of reinforcing learning at the institution is a rewarding experience
	To the students as previously described. [I could go on and on! Students in Chem 6 routinely outperform those who choose not to take Chem 6 in the General Chemistry course (Chem 12) for which Chem 6 is a 1-credit problem solving, optional companion course.] To the leaders, emotional and leadership skills growth. Also opportunities for scholarships and participation in dissemination activities. And many intangible benefits. To the instructor, myriad opportunities for scholarly development and activity, including conference presentations, working as part of MID dissemination teams, and preparation of materials for potential publication

the total of 60 respondents' comments from those individuals who gave the lowest and highest institutionalization ratings.

Among the respondents giving the lowest ratings for the likelihood of institutionalization, only one of eight mentioned student learning as a major benefit. Among the respondents giving the highest ratings for institutionalization, all nine mentioned student performance, some in eloquent terms.

Half of those with the eight lowest ratings mentioned affective benefits such as gains in confidence, morale, and satisfaction. Four mentioned benefits to the leaders. One made an unqualified reference to student performance, and one other made reference to improved learning among lower achieving students.

Not only did all of those with the higher expectations for institutionalization list student performance, they also gave learning particulars such as problem solving, content mastery, and conceptual understanding. As can be seen they also listed benefits in other areas. There is a strong sense in both the tone and content, that these respondents have experienced the success of PLTL with their students. It is, therefore not only perceived success that is an indicator of institutionalization, it is success in student learning that separates those who expect the program to persist from those who do not hold this expectation.

5.5.5 Correlations

In looking at correlations in general between ratings for expectations of institutionalization and ratings for other PLTL factors, the highest correlations were between institutionalization and workshops being adopted by other faculty (0.48), and between institutionalization and the belief that PLTL helps student learning (0.47). These are moderate correlations, but they are consistent with the other findings from the surveys, interviews, and site visits.

5.6 Summary

The data collected through surveys, interviews, observations, and detailed analysis—over a 10-year period—provided us with the opportunity to develop and to test five hypotheses about the likelihood of the program's persistence. We used Roger's definition of sustainability and institutionalization, namely that it is the continuation of a program after the period of initial dissemination and implementation. We find the following as good indicators that PLTL will continue at an institution and suggest them as components needed if any educational program is to endure.

1. Implementation according to the model and on-going fidelity to the model
2. Administrative support and funding
3. Perceived success, particularly in student learning
4. Fit with the institution's mission and practice
5. A core group of committed faculty

Chapter 6
Study of Former Workshop Leaders

From the earliest days of the PLTL project it was evident that leading workshops was of value to the peer leaders in a variety of ways. As is clear throughout this report, we had gathered data through surveys, interviews, focus groups, and journals (Dreyfuss & Gosser, 2006) that confirmed the benefits to the leaders. But we were also interested in the long-term impact of the program on former leaders. With a 10-year history the project was ripe for such a study. A survey was developed for a pilot study, revised, and a national study of former leaders was undertaken.

6.1 Pilot Study of Former Leaders

During 2001–02 a pilot survey was conducted with former PLTL leaders from St. Xavier University in Chicago where Varma-Nelson was a professor (Gafney & Varma-Nelson, 2002). Of 22 students who were contacted, 16 completed and returned the surveys. Most of the respondents were in their first position after college: three were employed as nurses; two were teaching in secondary school, one in community college; four were employed in science related industries; one was with a financial institution; two were medical students, one was working on a master's in teaching; two were still undergraduates.

The students in this pilot study were generally not science majors and many entered the course with anxiety about their abilities. The following comments illustrate the confidence gained by workshop leaders.

> *I entered St. Xavier as a nursing major. I knew I wanted to be a part of the medical field, but from past experiences in some scientific courses, especially chemistry, I was afraid I would not succeed. I entered the course with fear and uncertainty. I had an excellent teacher, who gave me confidence and I began to excel in the class. She encouraged me to be a workshop leader and my fear of chemistry was gone. I enjoyed the subject so much that I changed my major to biology. As a leader, I gained the knowledge and confidence I needed to pursue a career in pharmacy. During many medicinal chemistry courses in pharmacy school, I became known as the group leader.*

The former leaders in the pilot study gave high ratings to their experience as leaders and as students participating in workshops. With regard to impact on learning, the

L. Gafney, P. Varma-Nelson, *Peer-Led Team Learning: Evaluation, Dissemination, and Institutionalization of a College Level Initiative*,

highest ratings were given to: acting as workshop leaders, independent study of assigned work, participating as leaders in workshops, and attending lectures.

Table 6.1 shows how students rated the impact of various activities on their learning. This method is adapted from the Student Assessment of Learning Gains (SALG), developed by Seymour (http://www.wcer.wisc.edu/salgains/instructor/). Activities are listed in order of rankings, not in the order that they appeared on the survey. Rankings are based on the means for each item. Instructions to respondents were: For each of the following, please indicate the degree of impact on your learning when you were in college: 5 = very strong impact; 4 = strong impact; 3 = limited impact; 2 = very little impact; 1 = no impact.

Responses to open-ended items revealed that students viewed participating in workshops and acting as workshop leaders as: among their most productive learning experiences, increasing their confidence particularly in entering science-related careers, and making them more effective as they interacted with people in a wide range of situations. After the pilot study, we made minor revisions in the survey, and used it with a national population of former workshop leaders.

6.2 National Survey of Former Workshop Leaders

6.2.1 Design of the Study

The survey contains three sections. The first part asks the former leaders to rate the impact of different activities on their learning, with most items the same as those in Table 6.1. From this we obtained data about how the workshops compared, in the respondents' experience, with other teaching/learning activities.

Items for the second part of the survey were based on benefits previously noted by student leaders in interviews and questionnaires. These benefits were investigated with this larger population and over a longer timeframe. This part of the survey provided a platform for the third section, which asked open-ended questions

Table 6.1 Impact on learning at St. Xavier: ratings and ranking

	Mean	Rank	SD
Acting as a peer-leader for workshops	4.69	1	0.6
Independent study of assigned work	4.31	2	1.08
Participating as a student in peer-led workshops	4.27	3	0.7
Attending lectures	4.06	4	1.12
Working with a friend, study partner, or small group	3.81	5	0.91
Individual consultation with professors	3.53	6	0.83
Tutoring	3.31	7/8	1.2
Laboratory work	3.31	7/8	1.25
Independent projects, research, poster presentations	3.19	9	1.22
Off-campus meetings and conferences	2.94	10	1.29

about the impact of experiences on future decisions and events. Responses to the open-ended portion of the survey were sorted and coded according to categories that emerged. High levels of ratings for the second section would lead us to conclude that the qualitative material, provided in response to the open-ended items, was an articulation of generally positive experiences, not idiosyncratic or random responses.

To summarize, the study used a mixed method approach combining quantitative and qualitative methods to confirm and corroborate findings within a single study in what Creswell (2003) would call a concurrent triangulation strategy.

The survey was sent by mail or email to more than 570 former leaders from 11 institutions. About 12 percent of these were returned due to inaccurate addresses. The population was contacted three times with requests to complete the survey. Respondents completed the survey using a web link to the City University of New York (CUNY) Research Foundation. Data was organized by Victor Strozak of the CUNY Research Foundation. As with most surveys of this type, we cannot claim that we have a random sample. We have data from those who were interested enough to respond, and this interest may reflect more positive experiences. The respondents were reflective and insightful about the value of the workshops, and described a wide range of experiences.

Final results included 119 (about 24 percent of those actually contacted) completed surveys from respondents. These responses came from 11 institutions with 56 percent (67) from one institution. Of the respondents: 64 were female, 55 were male; 116 were undergraduates while they served as leaders, 3 were graduate students. Many of the respondents had worked as leaders during more than one year. The leaders served from 1995 to 2003; disciplines included: 31 in general chemistry; 84 in organic chemistry; 4 in biology; and 2 in mathematics; 107 (90 percent) of the respondents had participated in a leader training program.

Employment or educational status was reported as follows: 51 working in science-related fields; 5 working in non science-related fields; 9 teaching; 18 in medical school (including dental, podiatry, and veterinary); 10 in graduate school; 2 were not employed; 22 were still undergraduates; 2 no response.

6.2.2 Survey Outcomes: Learning Gains

For the first 13 items, students were asked to rate learning experiences on a 1–5 Likert scale. Mean scores greater than 4 are high for this type survey. In returned surveys, 56 percent of the respondents were from the University of Rochester. We are not able to give definite reasons for this outcome. In requesting responses, former students from Rochester were not treated differently from those attending other institutions, and although there were a greater number from Rochester in the pool, it was closer to 30 percent than 56 percent. We can conjecture that the former students from Rochester were more engaged with PLTL than those from other institutions, and responses tend to confirm this—as will be evident in the discussion. It may also be that addresses from the Rochester students were more accurate than those for students from other locations. In any case this outcome provided the opportunity for a comparison between Rochester and non-Rochester student responses.

Preliminary analysis revealed substantial differences between the Rochester and non-Rochester populations. The Rochester responses were generally consistent with those we had received from St. Xavier's in the pilot study. We had evidence that these two institutions had remained quite faithful to the recommended model in implementing PLTL. The survey results provided an opportunity to analyze differences based on leaders' experiences.

Table 6.2 compares mean ratings and rankings for Rochester and non-Rochester respondents. Activities are listed in order of rankings for the Rochester respondents. A look at rankings in Table 6.2 reveals that both groups (Rochester and non-Rochester) gave the highest scores to, "Acting as a peer leader for workshops." This experience was viewed as having the greatest impact on learning. Next, both groups gave high ratings to "Attending lectures," and, "Studying assigned work alone," indicating that these were generally conscientious college students, making good use of traditional teaching/learning methods. (Not all students participated in all of the activities listed, and so low rankings in some cases may be due to a lack of experience with this activity.)

The most important difference between the Rochester and non-Rochester students was in response to the item, "Participating as a student in peer-led workshops." The Rochester ranking was third, after, "Studying assigned work alone," with a rating of 3.9, considerably higher than the average, 3.38, for all ratings. (This item also ranked third in the pilot study of students at St. Xavier.) But the ranking of this item for non-Rochester respondents was 10, with a 3.5 rating, below the average for all ratings, 3.65.

We have to conclude that a number of workshop students—even those who went on to become peer leaders and who found that experience to be highly beneficial—

Table 6.2 Impact on learning: Rochester (R) and non-Rochester (n-R)

	Ratings		Rankings	
	R	n-R	R	n-R
Acting as a peer-leader for workshops	4.3	4.4	1	1
Studying assigned work along	4.2	4.1	2	4
Participating as a student in a peer-led workshop	3.9	3.5	3	10
Attending lectures	3.8	4.4	4	2
Acting as a recitation leader	3.7	2.9	5	12
Acting as a tutor	3.6	4.2	6	3
Working with a friend, study partner, or small group	3.5	3.6	7	9
Independent projects	3.2	3.7	8	7
Participating in a research project	3.2	4	9	5
Participating in a recitation led by a graduate student	2.9	2.9	10	11
Laboratory work	3.2	3.6	11	8
Individual consultation with professors	2.8	3.8	12	6
Receiving assistance from a tutor	2	2.4	13	13

did not find their student workshop experience to be substantially more important than many other academic activities. As mentioned above, we believe that the high ratings and positive experiences of the Rochester leaders, and of the St. Xavier students, are due to the care with which the program was implemented, particularly in the training and on-going work with the peer leaders.

Clearly, not all PLTL workshops are implemented in the same way or have the same impact on students. One respondent reported having had workshops in several courses, with differing levels of effectiveness.

> *I think the reason was because the department did not actually know how to utilize the workshops to their full potential. The problems were tough and the leaders were more used to being TA's and fulfilling that role. Also, they were often graduate students who did not understand the workshop model.*

Another commented (below) that the workshops introduced new and difficult material. The workshop model recommends that new material be introduced in lecture, not workshops. Since leaders are not experts in the subject matter, they are not in a position to teach new material. But they are good at learning and with training, they become skilled facilitators.

> *We were supposed to be using the whole discovery process. And if it had been introduced to me in more of a lecture format, it would have been a lot less confusing.*

These comments may help explain the modest rating given to workshop participation by some respondents. Like any educational initiative, the workshops can be implemented in ways that will have a positive impact, or they may be adapted in ways that diminish the impact. This is why it is important for each institution to monitor the implementation of PLTL and evaluate student experiences in their workshops

6.2.3 Survey Outcomes: Personal Benefits

The next set of items (Table 6.3) asked the former leaders to indicate agreement or disagreement with items about appreciations and skills derived from their leadership roles. For these items there were no significant differences between the mean ratings of Rochester (94.8) and non-Rochester (94.9) respondents.

Table 6.3 Percent agree or agree strongly for each item

Item	%
Acting as a workshop leader gave me an appreciation for the value of small group learning	95.8
As a workshop leader I gained an appreciation for different learning styles among students	91.6
As a workshop leader I gained confidence in presenting that was useful later on	95.8
As a workshop leader I gained confidence in working with a team that was useful later on	93.3
As a workshop leader I came to a deeper appreciation of what it takes to be a teacher	97.5

These are benefits that any institution of higher learning would be pleased to find in its graduates. This is particularly true of improved presentation skills, teamwork, and an appreciation of "what it takes to be a teacher." These percentages also provide a solid foundation for the open-ended items that follow. These highly positive responses about workshops cannot be attributed to respondents reporting what they thought the survey wanted. If that were the case they would have given much higher ratings to the value of attending workshops, as discussed above. It is, therefore, important to underscore the fact that while the non-Rochester respondents gave these very high ratings to their experiences as peer leaders, they did not, on average, give similarly high ratings to their experiences as students in workshops.

6.2.4 *Open-Ended Items*

The survey contained six open-ended items asking the former leaders to report the effect of the workshop experience on different aspects of their learning, personal growth, academic and career decisions, and overall benefits. In these open-ended items respondents selected areas to emphasize. They were not led or given topics from which to choose. Responses were analyzed and coded.

6.2.4.1 Learning

How did being a workshop leader affect your content learning and your ability to become a better learner?

- *Knowledge and problem solving.* Forty-three percent (51) reported that being a leader had provided a more thorough knowledge of the discipline and made them better problem solvers. Respondents said that the experience brought up questions and forced them to think about areas they had not previously considered. "It was the first time I realized how many gaps there were in my understanding of chemistry." One leader commented that the experience fostered repetition of material "in a seemingly non-repetitive way," and also allowed the leader, "to see different viewpoints and perspectives through the questions of others." Another said that this was particularly true because, "students asked questions that I would not have asked." Still others, following the same thought said that going through the material a second time makes it possible to, "appreciate the subject as a whole."

> *I truly believe that teaching someone else what I have learned recently is the best way for me to solidify my own understanding of a subject and improve both the integration of new knowledge with older knowledge and retrieval of information. As a workshop leader I developed the skills necessary to learn independently through the production of lessons (which is what I do often now as a teaching assistant in graduate student) and to learn through discussions—it often amazed me how much I could learn from the students in my workshops.*

- *Working in a group*. Twenty-six percent (31) commented that the group activities provided an understanding of different approaches, learning styles, and methods of problem solving. An intern in medicine reflected that, "Discussion of how to approach a problem as a group is fundamental to medicine, and being on both ends of the workshop spectrum certainly turned out to be training in this regard." Another said that, "By being a workshop leader I was able to acknowledge the importance of working with other students in coming up with a better solution."
- *Self understanding*. Twenty-three percent (27) discussed an increased awareness of their own approach to study and learning. A respondent wrote, "One of the things I gained from workshop leadership was getting to know my weaknesses and strengths. Knowing that definitely helps me to learn better using my strengths." Others reported that this self knowledge is closely allied to adaptability in learning. A graduate student said, "Because I was a workshop leader and was exposed to other learning styles and methods, I was able to adapt easily to a new learning environment."

6.2.4.2 People Skills

How did your experience as a workshop leader influence the way you interact with people?

- *Skills in teamwork, leadership, and presentation*. Forty-four percent (52) commented on enhanced skills in working with others, particularly in presenting material or leading a group.

> *I felt good that I was able to help other students and share my knowledge and skills with others. I also learned that communication is an art that has to be adjusted for each individual. I was more successful in dealing with different types of people after this experience.*

- *Confidence and patience*. Twenty-eight percent (33) reported increased confidence, comfort, or patience in working with people, particularly in teaching/learning situations. One student expanded on the quality of patience in a working situation.

> *It (acting as a workshop leader) gave me some patience in trying to find the different ways each person learns. I've found that useful in my work, in acknowledging that sometimes when people don't understand a concept, it's simply because they don't understand the way it's being presented to them. I try to be very clear on my expectations of my staff, and if I'm not, I just explain the concept from a different angle.*

- *Insights into differences and commonalities*. Thirty-two percent (38) described a new appreciation for differences among people, particularly in how they learn or understand new material. One leader commented, as follows.

> *Being a workshop leader helped me to understand that everyone interacts differently with people. It is important to not necessarily change how people act, but to work with their current personality type and show them how they can contribute to the group. It also made me understand that there can be different ways to solve one problem, and to be open to other people's ideas.*

6.2.4.3 Academic and Career Decisions

How did the experience of being a workshop leader influence your choice of future courses and your career direction? A majority of respondents said that their decisions did not change as a result of workshops. But a significant number reported an impact in the following areas.

- *Interest and appreciation for teaching.* Twenty-nine percent (34) said that they wanted to include some form of teaching in their careers. They found that they enjoyed explaining ideas and helping others learn.

  I'm one of the only TAs I know that actually loves teaching! Being a workshop leader helped me to overcome the fear involved in leading a classroom and instead to focus on helping each student to get something out of each workshop/discussion/lecture.

- *Confidence in studying science.* Ten percent (12) reported that the leadership experience had increased their own enjoyment and confidence in studying science. One said simply, "It helped me gain confidence in myself as a scientist." Another provided more detail.

  Being a workshop leader was far and away the best experience I had during my four years at college (academically and socially). It gave me great confidence in my abilities as a student and teacher and I performed extraordinarily well during that time and have excelled in many ways in life and academics since.

6.2.4.4 Workshop Pros and Cons

How did the workshops promote undergraduate learning? What were the best things about workshops? What are the worst things about workshops? These questions generated twice as much student leader response as any other question. Responses covered a wide range of areas. The following are direct quotes or paraphrases of student comments.

Among the best things about workshops, the former leaders cited the following:

- Workshops promote creative leadership and group process, teamwork. What the group knows is more than the cumulative knowledge of students individually.
- Students are more comfortable asking questions, even "silly" questions of other students than of professors.
- The workshops provide extra study time. Students are forced to keep up to date on material; to engage with the material.
- The workshops facilitate interaction among students, and expose students to other students' learning methods.
- Students are challenged to do more than they would on their own: Problems were very, very difficult and you needed a group of students to figure them out.
- Students learn how to organize study groups.
- Students saw others enjoy learning.
- There are many learning benefits to the leaders.

Among the worst things about workshops, the former leaders cited the following:

- Personality clashes among students can interfere with learning.
- It is difficult to engage students who are not interested in participating.
- Doing the problem sets did not leave time for other questions.
- The workshops involve a huge time commitment.
- Effective facilitation of the workshop model is hard to do well.
- Some students saw the sessions as more of a nuisance than a help. This led to apathy and discouraged the leader.
- Sometimes a strong student can dominate and the rest of those in the group just follow.
- It is difficult to deal with excellent students who learn best on their own.
- The workshops can become a crutch, taking away the motivation to study.

The items on the "best" list above describe major learning benefits that accrue to students generally, while the items on the "worst" list refer primarily to the behaviors or attitudes of a few students. There are however two items that refer to the time required and the fact that implementation is challenging. We agree that implementing the workshops is challenging and not easy to do well. The former leaders' perceptions of worst things point to the need for good weekly training sessions where workshop leaders' concerns can be addressed and solutions found to problems that arise. Methods from cooperative learning and from the PLTL literature (Dinan, 2004; Tien et al., 2004) can be useful in training leaders.

6.2.4.5 Recommendations

What recommendations would you make to professors who are teaching workshop courses? Respondents were enthusiastic about the value of workshops. Their suggestions covered a number of important areas. One comment nicely summarizes the benefits along several dimensions.

> *I would tell them that even though the investment in workshop courses is extensive at the beginning, the results that come from that investment are much more fruitful in terms of depth of student knowledge, and in the students' ability to connect with one another and build relationships, which ultimately lead to a better quality of life in college as well. It's definitely worth it!*

The following specific suggestions, from the survey, are in fact followed by most PLTL implementers. But their importance makes them worth noting here.

- Materials/workshop problems. *Write workshop problems with varying levels of difficulty, and problems that encourage different methods of approach.*
- Leader support. *Give the workshop leaders some type of support network or class while they are teaching. Most student leaders have little or no experience or training and it is helpful to have a group with which to discuss concerns and challenges and to give suggestions for teaching strategies.*

- Persistence. *Stick with it. I was in the entire introductory process and there were some bugs at the beginning, but it was never a negative experience for anyone. It's tough, but it will work well once the wrinkles are ironed out. Keep at it.*

6.3 Connections with Previous Studies

Astin (1984) proposed a theory of academic student development based on involvement. He distinguished his approach from theories based on the quality of courses available and taken; the presence of quality resources; and the institution's focus on individual needs. He defines involvement as the amount of physical and psychological energy that a student devotes to the academic experience. He identifies weaknesses in previous theories, and finds that involvement theory is simple but comprehensive and is positively related to virtually all of the factors associated with academic success and retention in college. Astin (1993) further elaborated his thinking about involvement, examining peer-groups at some length and concluding that the peer group is "the most potent source of influence on growth and development," during the college years.

Since Astin proposed his theory, others have conducted studies and added to the literature in this area. The most impressive study is that of Cress et al. (2001) in which data on 875 students from 10 institutions led to the conclusion that students involved in leadership activities, "showed growth in civic responsibility, leadership skills, multicultural awareness ... and personal and societal values."

PLTL workshops are implemented to improve learning by engaging students. This is the essence of the program—when workshops are implemented according to the model, students and leaders become more involved and their learning improves.

Results of the survey, tabulated and analyzed in this report, confirm previous anecdotal evidence that the workshop leaders reap significant, on-going benefits from their roles. Respondents report in overwhelming numbers that leading workshops reinforced the breadth and depth of their own learning, helped them develop personal qualities such as confidence and perseverance, and fostered a variety of presentation and team-related skills. These benefits were not associated with the nature of the institution—private or state, large or small, community college or four-year institution—but with the quality of workshop implementation.

The findings and analysis from our survey of former workshop leaders contribute to the growing body of literature on involvement theory, indicating that greater involvement and engagement, particularly in leaderships activities during college, leads to personal and professional benefits in the years following college.

Chapter 7
Impact on Minority Students and Women

Peer-Led Team Learning was developed as a way of involving all students more actively in their own learning, thus increasing their comprehension, problem-solving skills, and ability to work on teams. In this chapter we will discuss the impact of PLTL on two populations: under-represented minorities and women.

7.1 Academic Needs of Under-Represented Minority Students and PLTL

The educational shortfall of African American and Hispanic students in the United States has been well documented. Although African-American and Hispanic students comprise about 30 percent of the population, they earn only 16.4 percent of bachelors' degrees, 11 percent of masters' degrees and 5.8 percent of doctoral degrees in STEM disciplines (NSF, 2004). These figures are an improvement over 1979, for example, when Black and Hispanic students earned 6.2 percent of the bachelor's degrees (Miller, 1995, p. 39). But the numbers still stand as an indictment of the American educational system and our society in general. Under-represented minority students are often the first in their families to attend college, and lack appropriate role models at home. In addition, as the product of urban school systems, they are often academically under-prepared (Bottoms et al., 2001).

Miller (1995) offers three reasons for supporting efforts to remedy the situation and lessen the gap between the number of degrees earned by majority and by under-represented minority students at the college level. First, the achievement of all students is necessary for the productivity and competitiveness of the entire nation. Second, hard-won civil rights benefits will be maximized only if minority students gain an educational background comparable to that enjoyed by others. Finally, the maintenance of a harmonious and ethical society requires that minority students achieve at a rate equal to the population at large.

These reasons are widely accepted. But transforming words into practices that offset the handicaps of poverty, deprivation, and inadequate early schooling is not easy. Halpern (1994) pointed out an important issue in higher education, stating,

L. Gafney, P. Varma-Nelson, *Peer-Led Team Learning: Evaluation, Dissemination, and Institutionalization of a College Level Initiative*,

"The enrollment of large numbers of students from minority groups that have been traditionally underrepresented means that colleges have to respond with changes in the way they teach and in what they teach" (p. 2–3). Initiatives achieving some success at the college level have included: remediation particularly through not-for-credit courses (Kimbrough & Harper, 2006, p. 192), support services such as tutoring and workshops outside the course structure (Bonsangue & Drew, 1992), and the general improvement of teaching and learning through changes in existing courses. PLTL is located in this third area, that is the program aims to improve the academic performance of all students, not through remedial activities or tutoring/workshops outside the course of studies, but by modifying the course itself, adding workshops that expand the processes through which learning takes place.

The needs of minority students are met when courses are improved with more active learning and increased opportunities for student interaction and initiative (Seymour & Hewitt, 1997). When students are able to ask questions, engage in scientific discourse, and become fluent in the language of the discipline, they take ownership of their learning in new ways. Before addressing the special value of PLTL in this regard, we will briefly explore the attitudes and expectations of minority students.

There are many reasons for the under performance of African American and Hispanic students. Claude Steele (1999, 2003), investigating stereotypes, found that the very mention of race prior to test-taking had a negative effect on the performance of African American students. Groups of Black students who were told that Black students did not generally do as well as White students did in fact perform at lower levels than those in matched control groups that did not receive this information. Interestingly, White students told that in general White students did not achieve as well as Asians also did not perform as well as those in matched control groups. So the effect of negative stereotypes can be detrimental to academic progress for students in general. A high level of anxiety about failure inhibits both performance and motivation in students, no matter what their race or ethnicity.

A monograph by Landis (2005) describes the lack of progress of African American and Hispanic undergraduates in engineering in recent years. His hypothesis for lack of progress by minority students is, "that the primary barrier to minority student success at predominantly white institutions is the diminished quality of the learning environment resulting from: ethnic isolation, lack of peer support, lack of role models, and low faculty expectations" (p. 27). He then discusses three possible paths in facing the situation: inaction; ineffective action; effective action. The monograph then demonstrates the effectiveness of a collaborative learning environment, with impressive persistence statistics from two California universities.

In addition to the data supporting gains in academic performance through PLTL, we have evidence that the program addresses the social deficiencies encountered by minority students, described above.

- *Ethnic isolation and peer support.* If a large class contains a relatively small number of minority students, workshops can be planned so that minority students are not isolated. They are thus more likely to find a support group within the workshop.

- *Role models*. PLTL programs have selected significant numbers of minority students as peer leaders, wherever the program has been implemented. These young men and women become models to the next class demonstrating that it is "cool" to be a leader, helping others learn. In addition, these students become examples of leadership to majority as well as minority students.
- *Faculty expectations*. Capable minority workshop leaders demonstrate to faculty that minority students can accept increased academic responsibility and succeed in leadership roles. They also form a bridge between faculty and students, and help shape a positive, non-threatening learning environment.

As was mentioned earlier in this report, PLTL was inspired by the work of Uri Treisman (1985, 1992) who noted that the poor performance of black students at Berkeley was due more to academic isolation than to weaknesses in their earlier education. In imitation of what was seen to work for Asian students, he assisted African American students to form study groups and their academic performance improved substantially.

Although PLTL was not devised primarily for underrepresented minority students, there was, from the start, special interest in their success. A study at the University of Rochester (Tien et al., 2002) compared quality grades (ABC) for students over several years before and after the introduction of PLTL workshops in organic chemistry. Overall percentage of quality grades increased from 66 percent (n = 942) to 77 percent (n = 1,215). Percentage of underrepresented minority students earning quality grades increased from 47 percent (n = 85) to 58 percent (n = 158). These studies provide a solid foundation for the hypothesis that PLTL is particularly beneficial to minority students, with an increase of 23 percent (on a base of 47%) compared with an overall increase of 16 percent (on a base of 66%).

The City College of New York maintains an enrollment in chemistry classes that is more than 50 percent African American or Hispanic. The percentage of students earning quality grades (ABC) increased from a historical average of 38–58 percent with the introduction of Peer-Led Team Learning. In the earliest focus groups of 1994, discussion turned to mistakes and their role in learning. This discussion relates very much to the workshop benefits for minority students. The following is from an evaluation report written after those focus groups. At least half of the participants in general and of those quoted were minority students.

> *The importance of mistakes came up unexpectedly in the first focus group. One of the leaders said that the workshops gave students "the chance to make a lot of little mistakes." These mistakes in turn helped "make connections inside the brain." It seemed that this expressed an important truth. In the other groups the discussion was directed into this area, with further confirmation of the insight.*
>
> *Students said that fear of making mistakes was constricting. But in the workshops, "saying or doing something stupid is okay." When fear was diminished they were free to try out different ideas, to see where they led, to see what worked.*
>
> *Students described how workshop leaders sometimes made mistakes. They, the students, were not afraid to challenge the leaders, and to argue their points. (They said that if the professor made a mistake, they would think he/she was right and not offer a challenge.)*

One leader thought that after arguing a point of learning, "the conversation will stay in their heads longer."

The leaders felt strongly that mistakes meant much less in a workshop than in the lecture: "I'm not a teacher." "I won't remember." "If they get it wrong, it's not a big deal."

While most college students are afraid of making mistakes, it is important to recall that most of those in these groups, students and workshop leaders, were minority students with the added concern, discussed above, that a mistake would confirm the image and negative stereotype that they were less-able students. Student comments repeatedly affirm that the workshops provide a very important benefit as a location in which minority students, often with minority student leaders, can try out their ideas and test their preliminary understanding with considerably less fear and anxiety than they might otherwise have. In addition, students for whom English is a second language, can practice using English in the discipline as they participate in workshops (Varma-Nelson, 2006)

Bruffee (1999) recounts discussions in a course supporting peer tutors in collaborative learning. These tutors reported that some students were failing to make progress not simply because they were inadequately prepared, but because, "they appeared to be intellectually paralyzed" (p. 102). Based on the relationships with tutors these students were often able to overcome their anxiety, "discovering knowledge they did not know they had." In addition, working with tutors helped students overcome anti-intellectual tendencies.

7.1.1 Historically Black Colleges and Universities

PLTL has been adopted by several historically black institutions with positive results. Clark-Atlanta University in Atlanta was a participant in one of the Adopt and Adapt grants. This institution has been using workshops in general and organic chemistry for about 10 years, as of this writing. Clark-Atlanta has noted significant increases in student scores for the organic section of a comprehensive exam required of chemistry majors. Since PLTL has been the only curricular change, the increase is attributed to the PLTL initiative.

At Morehouse College in Atlanta, PLTL was introduced in general chemistry through a WPA grant. After two years, the grantee reported that PLTL had been institutionalized in chemistry, and was either piloted or had been introduced on a course-wide basis in biology, physics, mathematics, and psychology. In addition, the business and computer science departments had expressed interest. After two years of implementation, the success rate (percent of ABC grades) for general chemistry had increased by 10 percentage points, from a previous average of 52.2–62.5 percent. The use of PLTL in the other courses continued, although with different modes of implementation.

Howard University with a student body that is 77 percent African American and 8 percent Hispanic used a WPA grant to introduce PLTL into two sections of Chemistry for Health Sciences. The WPA report describes areas of success as follows:

- Grades improved. In the first semester of PLTL implementation, 74 percent of students in PLTL classes (n = 31) achieved quality grades (ABC) compared to 62 percent of those in the non-PLTL classes (n = 55). In the second semester it was 89 percent of those in PLTL (n = 46) compared to 67 percent in non-PLTL (n = 33).
- Many students who were reluctant to participate in the fall of 2004, participated in the spring of 2005.
- Thirteen capable peer leaders were identified among interested students, although not all could be used.

7.1.2 Northwestern University

Another study of the impact of peer-led workshops on minority students was conducted at Northwestern University (Born et al., 2002). (Although not associated with PLTL grants or founding group, the two-hour weekly peer-led workshops with trained facilitators are essentially the same as those developed by PLTL.) This was a two-year study of the performance of undergraduate minority and majority biology students. Volunteers were randomly assigned to workshops or to matched control groups. Due to small numbers it was not possible to have an assigned control group of minority students as part of the study, so an historic control group was used. One of the most positive outcomes of the study was that minority students showed a pattern of increasing performance on tests, as the semester progressed, compared with the historic control group of minority students which showed a pattern of decreasing performance. In the historic control group (n = 21), only 24 percent of the minority students remained in the course through the first quarter, and they earned grades of D or F. Of the minority group in the study (n = 25), none withdrew and none earned a D or F. Of the four minority students not in the workshop group during the time of the study, one completed the course successfully, two dropped the course after failing the first test, and one completed the course with a D. (Because of the small number and lack of matching with the workshop group, these were not considered a control group.) Majority volunteers randomly assigned to the workshop groups significantly outperformed those assigned to the control group (n = 60, $p < 0.05$).

The Northwestern study also used psychological scales to look at interest and anxiety regarding academics. The study found that those volunteering for workshops demonstrated greater interest in biology and reported that the course was more important to their futures than did those who did not volunteer. The volunteer group also registered higher in anxiety than the control group. Performance of control-group volunteers was negatively associated with increased self-reported anxiety, and so the report conjectures that those volunteers for the workshops may have been at increased risk academically due to their anxiety. (In other studies heightened anxiety is associated with poorer academic performance.) It may also be that the heightened interest and perceived importance of the course, along with the workshop experience, offset the anxiety.

7.1.3 Commuter and Community College Students

There are two other categories of college students that have a particularly large population of minorities: commuters and community colleges students. The large majority of Black and Hispanic students live in cities. Many of these students do not have the resources or may not wish to board at college, and many also choose, for economic and other reasons, to begin their post-secondary education at community colleges. Those enrolled in urban institutions and community colleges generally commute to school.

Students who commute are, almost by definition, less engaged in college life than those who live on campus. They tend to miss many of the informal encounters that take place between classes, on evenings, and weekends. Although commuters may join clubs, teams, and study groups even these activities must be carefully planned, and students usually head for home or jobs, soon after the activity time.

The following comments from a faculty phone interview with a professor from Miami-Dade Community College describes the evolution of workshops at a community college and the struggle to involve students. Miami-Dade's overall enrollment is approximately 70 percent Hispanic and 10 percent African American and this is reflected in general chemistry classes. Title 5, mentioned in the interview, is a federally funded program for minority serving institutions. The instructor quoted below also mentioned that the Title 5 programs tended to attract students with academic needs, but the performance of these students improved when the elements of the workshop were finally organized in a way that would insure their participation. The following process took place one semester at a time.

> *It started out with 'Isn't this a great idea and won't you please do it.' And one or two guys showed up. Then it became, 'I really mean it this time and I'm going to give you extra credit,' and everybody said, 'Oh fantastic,' and they came in the beginning and after a while they left because they found they might not do well anyway and other things were keeping them from putting their energy into the course. Then I said okay fine I'm making it mandatory, it's part of your grade, and decided to make it 12%. And they showed up. They said it was wonderful, but they weren't really doing the problems so I said, 'Okay now it's mandatory and you are taking a quiz.'*
>
> *Then the Title 5 grant came along and I was able to formalize it to what it is now. It is part of the course, absolutely mandatory, if you can't do it, go to another section—and there is a quiz. So now we have a captive audience that seems to be willing to show up. The results are a little bit hard to judge. But in the Title 5 section we have better retention. If you have better retention, you are doing better overall.*

PLTL classes at this community college demonstrated a success rate (ABC grades) went from 63 to 80 percent for chemistry; from 62 to 87 percent for biology; and from 69 to 81 percent for math.

Success stories similar to those described above have occurred at a number of community colleges and commuter schools. The following based on a discussion with peer leaders at San Jose community college reveals some of the dynamics and needs at these institutions. Students in the discussion were about 50 percent minority. The issues are related to under-prepared and busy urban students in general.

> *Student leaders described a number of factors involved in workshop success. First, many students are under-prepared, do not grasp much of what is presented in the lecture and need both answers to questions and group study to help them learn the required material. Second, leaders reported that tutors in the tutoring center are not sufficiently acquainted with the particulars of a given science or math text or method to comfortably guide students who drop in. Third, with outside responsibilities, many students do not devote sufficient study time, but the workshop forces them to attend to the material they need to study. The workshops provide extra learning time and a transition between lecture and individual study. Finally, after students represent their difficulties to peer leaders, they—the leaders—can in turn bring the issues to the attention of the professors.*

Minority students attending urban commuter schools and community colleges face a number of obstacles. The reasons for their limited engagement in studies is not simply the fact that they commute. They must often attend to responsibilities outside of school—work and family are the most common.

Kuh (2006) describes the results of a study matching engagement, assessed in a survey of 11,000 college students, with grades and persistence. Findings were that underserved students benefited more from educationally effective practices than mainstream students, in earning higher grades and persisting to the second year of college. This is important information for PLTL in so far as it adds a statistical foundation to the insights gained through experience, that minority students often reap greater benefits than majority students from course-related workshops.

7.2 Women and PLTL

The dramatic gains achieved by women in higher education during the past generation did not happen by chance, nor were they due to a few isolated programs or even to large-scale government intervention. Rather, as Whitehead (2006) points out, women have made dramatic advances, "because they are the beneficiaries of a sweeping popular reform movement that began more than 30 years ago and continues into the present" (p. 7).

Parents, mothers in particular, began to raise their daughters to think about careers and professions, and the education needed to achieve their goals in these areas. At the same time, reforms have taken place at every level of the educational systems. Young women have been particularly encouraged to engage in science and mathematics. Clubs, scholarships, mentoring, and group support have been made available—along with funding—from the earliest grades through college and university.

All of these activities, expectations, and opportunities have paid off. The National Science Foundation website reports that beginning in 2000, women earned more S&E (Science and Engineering) bachelor's degrees than men. The number of S&E bachelor's degrees awarded to women has increased every year since 1966 (excluding 1988), when it was about 50,000, reaching 227,813 in 2004. The number of bachelor's degrees in S&E awarded to men has fluctuated around 200,000 from 1976 to about 2001 and increased since then, reaching 224,525 in 2004.

But in some areas, the enrollments of women remain low. The number enrolled in graduate programs in computer science has declined in recent years, and the dot-com collapse impacted females to a greater extent than males. The percentages of women obtaining bachelor's degrees in computer science has declined from a high of 37 percent in 1983 to 26 percent in 1997. This decline has continued and the current figure may be under 20 percent.

Women in college have been quick to embrace workshops and to emerge as leaders. Fifty-five percent of the returns for the survey of leaders described in Chapter 6 were women. Our observations of many workshops and discussions with groups of leaders also suggest that the majority of PLTL leaders have been women—although we do not have overall statistics for this. Faculty interested in fostering the progress of women in areas of science in which they remain under-represented have recognized PLTL as a supporting activity.

In order to attract and engage young women and minority students in computer science, a consortium of seven colleges received a National Science Foundation collaborative grant starting in the 2005–06 academic year, and PLTL was introduced as a key strategy. A summary of the strategy is stated in the following excerpt from the program documents.

> *Incoming freshmen in targeted under-represented groups with strong math/science backgrounds will be actively recruited to enroll in an introductory CS course that includes a special additional discussion section. That section will meet for two hours each week during which the students will work in small groups on challenging problems designed to help them gain a thorough and in-depth understanding of the class material, and to increase their enthusiasm for Computer Science. The discussion sections will be run by outstanding undergraduates (at least half of whom will be women and/or minorities), who in turn will receive extensive supervision and training in how to facilitate group learning.*

The use of peer-led workshops was selected because the approach is relatively inexpensive to implement and does not require departmental reorganization. To test the approach, different types of institutions were selected for participation. These included large state schools, research institutions, and small liberal arts colleges. At this writing, the collaboration has completed just one year in which the project was piloted. Nevertheless, evaluations were conducted and initial results were positive.

Peer-Led Team Learning is relevant to this discussion because participation in workshops creates experiences and a climate through which students cooperate with one another regardless of gender. Participation as students and as leaders prepares students, female and male to see themselves as equals, to expect and to demonstrate equal treatment. As reported in Chapter 6, these experiences are carrying over into graduate school and the job market, where women are increasingly judged on the basis of performance rather than gender.

The University of Rochester study that was referenced above (Tien et al., 2002) revealed the disproportionate benefit of workshops to female students. In the control group, males outperformed females in attaining quality grades 70–62 percent, but females outperformed males in the PLTL group, 78–76 percent. The women's success increased from 62 percent in the historical group to 78 percent with PLTL. The men's success rate increased from 70 to 76 percent.

After PLTL was introduced in the biochemistry course for allied health professionals at St. Xavier University in Chicago scores on the standardized ACS exam increased from 1 percent above the national average (N = 127, two years) to 8 percent above the national average (N = 146, five years). Enrollment in this course was on average more than 95 percent women. These statistics provide further evidence that the PTTL workshops are particularly helpful to female students.

Similar results occurred at Northeastern Illinois with the introduction of PLTL in precalculus and calculus courses. In this case workshops were added to a preexisting problem-solving seminar. The first result was that participation in the seminars increased from a historic average of 3–8 percent to 13.5 percent (N = 1,423) for eight PLTL courses over four semesters. Participation of women in the seminars was 60 percent, while enrollment in the courses averaged 51 percent women.

7.3 Summary

We have found that under-represented minority students, primarily African American and Hispanic, have benefited academically from Peer-Led Team Learning. We have identified the following issues and corresponding PLTL benefits.

- *Weak academic backgrounds*. PLTL provides additional facilitated problem solving and study time to reinforce textbook and lecture. Implementation studies reveal the success of minority students, outperforming control groups.
- *First generation going to college*. PLTL offers support, motivation, and role models by helping minority students work together and by providing minority workshop leaders.
- *Commuter and community college students*. PLTL offers an out-of-class group and context providing students with stronger links to particular classes and to the institution.
- *Under-representation*. PLTL can offer a group with which minority students can identify if, as is sometimes the case, they feel left out and uninvolved with a class in which there are very few students "like them."

The situation for women in science is different but has some similarities with that of minority students. There are relatively small numbers of women in physics, computer science, areas of math and some other sub-disciplines. With small numbers of women there is sometimes a male-oriented culture in a department. In this situation, women need special support. As noted in the chapter certain grants and programs have recruited women and used PLTL as an effective teaching/learning strategy to reinforce learning and provide a locus for networking and offering mutual support.

Chapter 8
PLTL and the Goals of Higher Education

Introductory science courses are frequently viewed as hurdles to be cleared on the way to the more important and more interesting courses. Faculty may see themselves as gatekeepers, identifying students who are capable and allowing them access to the advanced courses. Tobias & Raphael (1997) point out that a professor of introductory physics may view the course as covering important concepts in motion, optics, mechanics, and so on. But for students, the course consists of problem sets, lab reports, quizzes, and a final exam. They go on to say that this difference might not matter except for the fact that the two perspectives can be in conflict. When a student has to make decisions about how to spend time and energy, or where to put in extra work, the grade value involved will probably play a larger role than intrinsic interest in the material.

Peer-Led Team Learning can bridge these two perspectives by increasing student interest and motivation and improving learning while not concentrating on grades. Through exploratory activities and group work, students become more involved with the material. As they learn, their confidence grows, and this contributes to better grades—as students think less about them.

8.1 College as Preparation for Careers

Most students will change careers several times during their lives, and if they stay in the same career, their work will change. They will need to be self directed lifelong learners, to think creatively, to recognize patterns and have the ability to see "the big picture." Students need good written and oral communication skills and the ability to work with people from different cultural backgrounds (Pink, 2005).

Maxfield (2001) reports the results of a survey at Allied Signal company, which found that when considering both importance to industry and the need for better preparation, the greatest personnel needs were in communication, oral presentation, and working as part of a team. At a recent "Workshop on Research Evidence Related to Future Skill Demands" organized by The National Research Council (NRC), Houston

L. Gafney, P. Varma-Nelson, *Peer-Led Team Learning: Evaluation, Dissemination, and Institutionalization of a College Level Initiative*,

(2007) reported an increased demand for workers with the following six basic competencies:

1. *Creative problem-solving*: employing unique analyses and generating new, innovative solutions to problems; integrating seemingly unrelated information and developing creative solutions; entertaining possibilities others may miss.
2. *Complex communication skills*: knowing the appropriate channels for getting things done; developing and maintaining partnerships; effective negotiating and persuasion skills, influencing without authority; team-building skills; communication as appropriate for different messages or situations (e.g., when to pick up the phone vs. sending an email).
3. *Adaptability*: ability and willingness to cope with uncertain, new, and rapidly changing conditions; flexibility in approaches to work.
4. *Self-management*: ability to work remotely in virtual teams, ability to work autonomously; self motivating and self monitoring.
5. *Self development*: willingness and ability to acquire new information and skills related to work.
6. *Systems thinking*: the ability to understand how an entire system works, how an action, change, or malfunction in one part of the system affects the rest of the system; adopting the big picture perspective on work.

It is clear that we cannot continue to employ strategies for teaching that were used by previous generations. If students are to acquire the skills that are outlined above change is required. While no single teaching method can provide all the skills students need to be successful, student-centered pedagogies that promote active learning will be essential.

PLTL workshops provide an opportunity for students to do more than listen in class and give back on tests. They are able to construct new knowledge for themselves, similar to the way scientists generate new knowledge in research teams. Varma-Nelson and Coppola (2004) point out that the structure of peer-led groups is similar to that of the research group which has been so successful in educating graduate students in the sciences. Students in PLTL groups, like researchers, try out their ideas on each other; discussing, refining and revising their positions, listening to new evidence, finding out where they went wrong and where they can suggest new approaches to colleagues, so that they arrive at an understanding deeper than any had at the outset. These interactions promote cognitive, social, and epistemological development. Students learn; they interact with others in positive and professional ways, and they develop new understandings about how learning takes place.

In a peer-led learning environment the students have many opportunities to develop and maintain partnerships and learn new material through negotiation and persuasion. The peer leaders learn to "influence without authority."

While the workshop experience is not a panacea, a student who has been challenged in a group setting, who has cooperated with peers in solving problems, and who has been forced to consider the place of new data or to devise new models will be better equipped to function on a team in the workplace than one who has studied in isolation.

8.2 Changes in Teaching and Learning

A recent article, "The Tough Road to Better Science Teaching" (Brainard, 2007), describes the situation in college science teaching and the difficulties in implementing change, particularly at research universities which award 57 percent of the degrees in the science, math, and engineering disciplines. Professors are slow to change; departments find it difficult to agree on change; administrators cannot require change. And yet there is ample evidence that inquiry-based, student-centered learning succeeds where traditional lecture/memory methods do not.

Connolly and Millar (2006) point out that the first stage in instructional change depends on challenging or unfreezing assumptions about the effectiveness of one's behavior. This stage is followed by the exploration of alternative behaviors. Finally, there is the most difficult stage of attempting a new approach and dealing with feedback that may be mixed.

In previous sections of this study we have noted that PLTL is more likely to be successful when the workshop approach to teaching and learning is aligned with the educational priorities of the individual faculty, department, and institution. We can now discuss in greater detail what this fit means, particularly with regard to attitudes toward teaching and learning.

Fink (2003) discusses issues in higher education. On the positive side, he finds new understandings about how learning takes place, rapid advances in the production of knowledge, evolving expectations about how higher education should serve the individual, and the entry of technology into education. On the negative side, he notes the fact that large numbers of students are leaving college with an inadequate foundation—in their knowledge base, critical thinking skills, facility in writing, and even in human relations—for the tasks that await them, personally and professionally. He believes that the advances in pedagogical techniques can be used to address the current problems. A paradigm shift is necessary if this is to take place. Fink presents the characteristics of the old and new paradigms in table form (p. 19), part of which is adapted below.

Table 8.1 Old and new paradigms for college teaching

	Old paradigm	New paradigm
Student	Passive vessel to be filled by faculty's knowledge	Active constructor and discoverer of knowledge
Student growth goals	Complete requirements, graduate with a major	Focus on lifelong learning in a broader system
Relationships	Impersonal relationships between faculty and students	Personal relationships between faculty and students
Context	Competitive, individualistic	Cooperative learning in classroom and on faculty teams
Power	Faculty holds and exercises power, authority and control	Power is shared among students and between students and faculty
Teaching assumption	Any expert can teach	Teaching is complex and requires training

Many faculty members who have witnessed the positive changes in teaching and learning that are achieved with PLTL agree with the shift and the new priorities indicated in Table 8.1. The table nicely summarizes an evolution in thinking, in attitudes, and in working—as one enters a learner-centered environment.

The following are strategies Fink cites that exemplify the paradigm shift discussed above, with examples of relevant programs used in conjunction with PLTL. These are only a few illustrations of the fact that PLTL instructors are often engaged in serious discussion and analysis of programs and pedagogies through which students take a more active part in their learning.

- *Writing to learn*. Various recent programs demonstrate that writing can enhance the quality of learning in all disciplines (Bean, 1996). At the University of New Hampshire, faculty introduced Calibrated Peer Review, a program using writing and peer assessment to enhance instruction, along with PLTL.
- *Small group learning*. When properly implemented, small groups can facilitate the problem-solving process, foster academic improvements, and increase cross-cultural awareness (Millis & Cottell, 1998). Small group learning is of course central to the PLTL method.
- *Assessment as learning*. The assessment of learning need not be simply a means of grading, but can contribute to learning if students are included in the process (Angelo & Cross, 1993). Faculty at Canisius College and other sites have used team-based tests, to supplement individual tests, with workshop courses.
- *Problem-based learning*. Imitating methods used in medical schools and other professional areas, students learn in ways that simulate real-world conditions (Duch et al., 2001). A professor at the University of Rochester used problem-based learning in conjunction with PLTL in a biochemistry course.
- In addition to the above, PLTL used distance learning at the City University of New York, interdisciplinary courses at Goucher College, and a number of other student-centered pedagogies in conjunction with workshops.

8.3 Faculty-Student Partnerships in Teaching and Learning

Faculty members who are reluctant to share teaching/learning responsibilities with undergraduates tend to be wary of PLTL. They do not want their discipline or themselves to be "misrepresented." In previous chapters we have described these views as barriers to be overcome if PLTL is to be successfully implemented. At this time we want to consider a more positive perspective on the situation. Peer-led team learning is part of a growing movement of faculty-student partnerships in higher education.

In a variety of ways the peer-leaders and their work with professors serve as a bridge between faculty and students. First, the peer-leaders meet weekly with the professor either as part of a teaching/learning course or in a leaders' meeting that is not part of a course. In both cases, there are discussions about content, learning, workshop strategies, group dynamics, student response to the program, and the like.

These sessions create a partnership because faculty and student leaders are engaged in a common enterprise beyond the classroom—improving the learning and academic performance of students.

This common effort is indeed a partnership because faculty look to the leaders for insights into student learning, for feedback about the effectiveness of various strategies, for assistance in reviewing, constructing and testing materials, and for information about topics that are particularly challenging to students. Some of this information is traditionally available to faculty through office-hour contacts with students. But the leaders provide another dimension to the communication because they can make suggestions about how to improve lectures, materials, and contact with students. Their weekly journals have also offered a rich source of insights into the progress of the program and the dynamics of student learning.

A second dimension of the partnership is found in leaders bringing the professor's priorities and perspective to students. They are in a position to say authoritatively, "This is what he meant," or "This is what she considers most important in the chapter." We have often witnessed faculty members in the weekly sessions with leaders asking about student response in the previous workshop and providing advice for the coming workshop in words such as, "If they are confused about this, use the following example." The bridge formed by the leaders definitely carries traffic in both directions.

Finally, the partnership between leaders and faculty serves as a model to students affirming that professors are approachable and concerned. Leaders often share the responsibility of selecting the next cohort of workshop leaders. Even students who are not selected as leaders benefit from the faculty/leader partnership through their workshop participation and the understanding that the professor is concerned about their learning and professional development.

Miller et al., (2001) provides an extensive review of faculty student partnerships in a collection of reports that describe 31 programs, citing a multitude of benefits derived from "student–assisted teaching." They include: positive affective outcomes, use of powerful role models, addressing the social nature of undergraduates, and increasing student responsibility for learning. These benefits taken together can lead to a renewed sense of connectedness between faculty and students.

This discussion, we believe, makes it clear that faculty-student partnerships should be viewed as positive resources that can strengthen higher education, making learning more effective, long-lasting, and personal. The participation of undergraduates in the education of their peers should not be viewed as a concession to remedial or financial requirements but as the use of a significant untapped resources that can facilitate learning (Gosser et al., 2001, p. 3).

Finally, regardless of their pedagogical philosophy, faculty members share a common goal—that their students' interest in science grow and that more qualified students decide to major in science. Astin (1993) based on a longitudinal study of more than 27,000 students found that a larger proportion of a student's peers studying science is a predictor of the likelihood that a student will persist as a science major and enter a science related career. In other words if your friends are in science, you are more likely to remain in science; and PLTL promotes friendships in ways that attending lecture does not. Studies such as those of Light (2001) and our own interviews have repeatedly found that science majors surrounded by academically like-minded peers are much more motivated and more likely to persist in science than those whose friends are predominantly not science majors.

8.4 Conclusions

With regard to new paradigms for teaching and learning, the essential point is that there has been a shift from the focus on teaching to a focus on learning. Although teaching and learning are opposite sides of the same coin, the question of focus is not insignificant. When professors plan courses, what are they thinking, and what do they do? Generally, they plan, perhaps very carefully, a series of lectures—with examples, references, connections, illustrations, etc. All of these are intended to provide the clearest possible presentation of the material. The assumption is that if the material is presented accurately and clearly, then learning should take place—as long as the student is responsive.

But does the professor think for long about exactly what the individual student or groups of students will be doing to create their own learning? The response generally would be that the students should pay attention during lecture, ask questions of themselves and others on material they do not understand, do the assignments, and prepare for tests. The problem is, this may not be enough.

We have shown in this chapter that many believe the shift in focus from teaching to learning, results in significantly different learning activities. From our vantage point, Peer-Led Team Learning is one. The focus in workshops is on learning. There is no teacher. The job of the leader is to facilitate learning. The task of the student is to learn and to help others learn, so that the whole is greater than the sum of the parts.

Based on our study we propose that a serious focus on learning should precede the introduction of peer-led workshops. It is those teachers unwilling to enter at least the first steps of the paradigm shift who will either not attempt to implement PLTL or will abandon it when problems arise.

Finally, we have found that the teaching/learning enterprise is much improved when faculty enter into partnership with students. As recent learners of the course material peer leaders have unique insights about issues related to learning, and they have a connection with other students that puts learning science in a social context. The subject matter becomes more approachable for students because it is modeled by one of their own.

Chapter 9
Issues in Implementation, Dissemination and Institutionalization

We have traced the evolution of the PLTL model of teaching and learning from a local setting to a national program through implementation, dissemination, and institutionalization. For each stage we have used theory and the experiences of many practitioners to explore and explain PLTL. In this chapter we summarize findings and discuss how the project can be further strengthened. We also review critical components for successful implementation and institutionalization of PLTL.

This study has found that adding peer-led workshops to a course requires a faculty with both flexibility and motivation. Flexibility is needed because workshops call for a shift in roles and expectations about how teaching and learning take place. Motivation is needed because introducing and maintaining workshops requires additional time and effort, beyond what is demanded for lectures. This is particularly true when new workshop materials are created and strategies for leader training must be learned and implemented. Some changes involve offices outside the department, such as the registrar and the learning assistance center.

9.1 Issues in Implementation

9.1.1 Critical Components

Identification of the critical components for successful implementation has been useful in the evaluation of PLTL programs across the country. The critical components were listed and explained in Chapter 2, and we have made frequent reference to them throughout. In summary, they are:

1. *Faculty involvement*. The faculty members teaching the course are closely involved with the workshops and the workshop leaders.
2. *Integral to the course*. The workshops are an essential feature of the course.
3. *Leader Selection and training*. The workshop leaders are carefully selected, well trained and closely supervised, with attention to knowledge of the discipline and teaching/learning techniques for small groups.

L. Gafney, P. Varma-Nelson, *Peer-Led Team Learning: Evaluation, Dissemination, and Institutionalization of a College Level Initiative*,

4. *Appropriate materials*. The workshop materials are challenging, intended to encourage active learning and are appropriate for groups.
5. *Appropriate organizational arrangements*. The particulars, including the size of the group, space, time, noise level, etc. are structured to promote group activity and learning.
6. *Administrative support*. Workshops are supported by the department and the institution as indicated by funding, recognition, and rewards.

The system of using six critical components as a guide to implement PLTL has been valuable for a number of reasons. First, it provides a clear description of what the method means: for those presenting it, for those interested in learning about it, and finally for those adopting the program. Second, the critical components offer a checklist for assessing the quality of implementation. We have found that problems in implementation are invariably due to weaknesses in one or more of these essential features. Identifying the problem area makes a solution more likely. Third, because the system clearly defines the program, it also allows for adaptation—across disciplines, types of courses, institutions, and in combination with other pedagogical methods.

The critical components also define the limits of adaptability for the PLTL model. For example, it is absolutely necessary that faculty be involved in the design of the program at each adopting institution. But design and implementation will proceed more smoothly if faculty are assisted in the preparation of materials, in scheduling, and with program administration. Similarly there is room for variation in the length of the workshop. While a length of two hours is recommended, we have found that workshops of 90 minutes are also effective in improving student performance. When workshops are much shorter, the gains in student performance are not realized. There are different models for the weekly training of leaders and the choice of a particular method depends on the size of the program and resources available at an institution. Without weekly training sessions and without faculty involvement in the training sessions, failure is virtually guaranteed. Lack of support from the dean and provost can also be problematic.

9.1.2 Changing Faculty and Student Expectations

In addition to the attention that must be paid to the critical components in implementing PLTL, faculty and student expectations about teaching and learning need to be addressed. Interviews, focus groups, and other evaluation data demonstrate that years in the classroom create ingrained thinking about educational procedures. This is true for faculty and students alike. Faculty have definite ideas about teaching and most of them involve faculty-delivered instruction as indicated in this quote from a participant at a PLTL dissemination workshop, "At this point I am not completely convinced that I should allow the direction of these small problem-solving groups to be handled simply by undergraduates who've had the general chemistry sequence (and done well)."

Students who have experienced only the "sage on the stage model" of instruction (King, 1993), expect to see the instructor standing on stage making a presentation. They like discussion, but particularly in the sciences students want information. Erickson & Strommer (1991) finds the following typical of freshmen chemistry students: "The teacher is good in his field. He encourages us to think. But he doesn't define things. Thinking is fine, but learning is what I'm here for" (p. 49). The PLTL model upsets comfort levels, shifts expectations about roles of faculty and students, and introduces a third population the peer leaders who form a bridge between the faculty and students. All three groups have responsibilities in making the transition successful. The roles of faculty, peer leaders and students should be clearly defined during the training of leaders and in early classes and workshops for students. The literature on cooperative and team learning is useful in this training.

Gafney, has observed more than 80 PLTL workshops and has seen that while the dynamics are generally positive with real learning and helpful exchanges taking place, there was room for more activities that reflect structured cooperative learning. As PLTL has grown over the past decade, evaluation data and the experiences of practitioners indicate that in addition to training faculty and leaders, the participating students should also receive training. Because the workshop is an unfamiliar environment, students need training in workshop dynamics in order to become familiar with the expectations surrounding PLTL.

Training students in workshop participation has been neglected. We recommend that all three groups—faculty, student, and peer leaders—receive a careful and thorough introduction to the model at the start of the program in order to ensure smooth implementation. Both faculty and peer leaders have a role in training the students. Faculty have the responsibility to inform students about their role (in workshops) during the lecture and the leader exercises this responsibility during the weekly workshops. Leaders should incorporate workshop time to reflect on how the group is doing as a team. Active involvement of all three groups can decrease the time it takes to create the appropriate learning environment.

In recent years, cooperative or team learning has been widely used in many different formats at the college level and as a result a great deal has been learned about making team learning successful. We recommend that PLTL practitioners incorporate Johnson's learning group (1991) ideas into the semester's overall activities, and Michaelsen & Knight's (2004) practical suggestions for creating effective group assignments and promoting social learning (p. 53).

An earlier report on cooperative learning (Felder & Brent, 1994) describes the careful implementation of team-based learning in five successive chemical engineering courses, with substantial gains in student performance. A variety of in-class two- and four-person exercises are explained; followed by a discussion of productive out-of-class team exercises. Finally, the report offers detailed answers to a number of typical faculty concerns, for example, "If I spend all this time in class on group exercises, I'll never get through." The report provides an excellent primer on how to get started with cooperative learning—with strategies that have had proven success.

Each of the three PLTL stakeholders—faculty, leaders, and students—can profit from increased attention to and training in cooperative learning and process skills.

This training would then be reinforced throughout the semester by both faculty and leaders. Three projects that incorporated such training are Lewis's modified peer-led guided inquiry (2004), Dinan's alternative to lecturing workshops in organic chemistry (2004) and Hanson's process workshops (2000).

By reviewing projects like these, PLTL can keep the door open to other relevant cooperative learning and group processing skills. Johnson et al., (1991) offers ideas for group meetings (6: 16) that include questions about the completion of preparatory assignments, and about what has been learned. Many leaders already use these techniques. An additional strategy would be to draw students into the discussion so that they ask and answer one another with these kinds of questions. Such exchanges are in fact a PLTL expectation, but they are often missing in practice.

There is a wealth of literature on cooperative and team learning. We suggest a few strategies that might benefit PLTL workshops: (1) Team exams. Dinan (2004) describes an approach in which a test is taken first by individuals, then by teams. The individual score outweighs the team score by four or five to one, but the exercise gets the team working together and reinforces learning. (2) Required teamwork. Mazur's conceptests (1997) require paired discussion because students must explain and discuss their responses. (3) Hard problems. Bradley et al. (2002) describe problems that none of the students could do alone, so teamwork is essential.

Feedback is a frequently mentioned group processing skill (Michaelsen 2004; Birmingham & McCord, 2004). Michaelsen states that feedback is most effective in learning and in team development when it is immediate, frequent and, "enables learners to clearly distinguish between good and bad choices, effective and ineffective strategies" (p. 33). Again, feedback to students is a common strategy in PLTL, but processes should be introduced so that students in the group as well as the leader use structured feedback to one another to strengthen both learning and team performance.

9.1.3 Faculty Involvement in Teaching/Learning Initiatives

This study has reinforced what is known about the profession of college teaching. In many ways faculty members have the best of the two worlds of security and independence. With tenure they have greater security than those in most other professions and they have a substantial voice in determining the arrangements of their work, particularly the courses they teach and scheduling their time. Beyond these requirements, college faculty select their areas of research and service. Faculty rewards in science departments are primarily centered around quantity of publications and presentations in the disciplines. Non-tenured teachers at a dissemination conference (Ames, Iowa, 1999) registered concern about whether adopting PLTL would be the best use of their time. Some were distressed to hear that student ratings of lectures sometimes declined after PLTL was adopted, because students valued the workshops and seemed to appreciate lectures less. These were common concerns among potential adopters. On occasion students did not value participation in workshops, and this resulted in negative comments on the end of the semester evaluations. These usually disappear with time but non-tenured faculty don't have much time.

We strongly encourage that those who are establishing guidelines and making decisions about tenure should look very closely at strategies for achieving appropriate student input when new approaches to teaching and learning are implemented. In addition, they should adopt methods of gathering data that take a holistic approach to student self-assessment, that is determining whether students find that the total program, in this case including workshops, leads to more effective learning.

Several factors can militate against the adoption of PLTL. First, the method requires time and effort to implement and maintain. This may interfere with time devoted to research or publishing, and even faculty members with tenure often object to interference with these critical areas of their professional lives and careers. Second, PLTL requires that faculty rethink the processes of teaching and learning. They often object, at least implicitly, to sharing responsibilities with undergraduates whose understanding of the discipline is incomplete. This attitude comes from lack of understanding of the role of the peer leader.

Throughout this study we have pointed out that there are answers to these objections. But they are answers that will only appeal to those who are receptive to new ideas. Faculty members who believe that PLTL will interfere with what is best for their careers or that undergraduate leaders may jeopardize learning are generally unwillingly or slow to change their views. Some faculty attended lectures about PLTL several times before actually making the change in their classes.

Perhaps the attitudes and beliefs of faculty can change with more serious discussion about teaching and learning. Responses to a National Education Association survey of school teachers (Rait, 1995) have implications at the college level. Considering factors that positively impacted their teaching, respondents gave the highest ratings to their own experience. Reflecting on their own activities to promote learning was viewed as the most significant catalyst for change. The next highest rating went to consultation with other teachers, and third to observations of other teachers. Consulting with experts and professional development were far down the list.

Based on numerous interviews and discussions, we find that college professors hold similar views regarding improvements in their performance. They tend to change their approaches based primarily on their own experiences and what they learn when consulting with colleagues about teaching and learning. These different influences do not work independently, but in concert. Although professors gain daily experience in teaching and learning, real reflection takes place when they must explain or listen to others explain the dynamics of the teaching/learning process.

The problem for college professors is, of course, that interactions with colleagues about teaching may be rare compared to interactions about their research, and actually seeing another professor in action may be even rarer. This we believe is an important missing piece that prevents greater forward movement in the adoption, improvement, and institutionalization of Peer-Led Team Learning and other initiatives. College professors rarely meet to discuss how they teach and how students respond to their teaching. Coppola (2007) points out that faculty are responsible for developing future scholars and this includes both classroom teaching and research. Therefore by extending the discussion among faculty from research to research-and-teaching we can tap into the intrinsic aspects of what it means to be a faculty member. It must include continual learning, with reflection on what it means to guide students in their intellectual progress.

Faculty development workshops provide a forum for promoting this discussion. Hutchings and Shulman (1999) conclude an article on the scholarship of teaching suggesting that institutional research offices should get into the mainstream of the university looking for answers to questions like: What are our students really learning? What do they understand deeply? What kinds of human beings are they becoming? If this were to happen, professors would take notice and be forced to reflect on their teaching.

We conclude this discussion of approaches to pedagogy by considering three teaching roles suggested by McKeachie & Svinicki (2006): expert, facilitator, and role model. Each role has its place, and should be used appropriately. When playing the first role, the professor lectures, has the answers, tries to impart knowledge. The second role focuses more on the learner, and the instructor (or peer leader) acts as a guide pointing the way but letting the student do more of the work. The professor as role model, in a way bridges the other two approaches. He or she, as researcher and explorer, demonstrates the processes of discovery. In PLTL, at its best, professors use each of these roles at different times. The peer leader also acts as facilitator and role model.

9.2 Findings About Dissemination

We have learned that dissemination of an educational innovation should not be treated simply as a marketing endeavor. PLTL practitioners encouraged others to use the method, but they also endeavored to make public their work and findings in quality journals and professional conferences. They were available for consultation so that others could introduce the approach in their courses and judge its value to their students. Thus the transportability of the method and justification for its use could be evaluated by others in their own environments. In this context of implementation and mentoring, new leaders emerged naturally and new ideas were developed by adopters. This process contributed to the evolution of the program. Local dissemination was required by the project's WPA grants, and grantees usually did much more than was expected. On the basis of mini-grant reports and site visits, Gafney found that the four-stage national dissemination plan developed by the project was well conceived and worked well in accomplishing the program's goals. These stages from the two perspectives of disseminator and implementer are shown in Table 9.1.

This model gave the project an overall mechanism for further growth. Professional conferences and publications, particularly in chemistry, and hundreds

Table 9.1 Dissemination stages

Stage	Disseminator	Implementer
1	Provide initial information	Gain basic understanding
2	Offer detailed understanding of how-to knowledge	Acquire the conceptual knowledge, skills, and motivation suitable for implementation
3	Provide technical and monetary assistance and mentoring	Implement workshops
4	Offer collegial support for emergence of leadership	Take professional responsibilities for further analysis and dissemination

of talks and workshops provided a forum for presenting and promoting the project. In addition, sponsoring organizations such as the NSF-funded Chautauqua Faculty Development Program, regional and national American Chemical Society meetings, and Gordon Research Conferences on education created an ongoing platform for the presentation of basic ideas and data about the project. As implementation grew, members of the original PLTL team were available as consultants and made visits to sites adopting PLTL workshops, discussing what had worked in their own experience, what adaptations might be useful, and how to respond to problems associated with new implementations.

While not all new adopters received grants from the national PLTL project for implementing workshops in their own courses, the most successful dissemination strategy was the use of the WPA (Workshop Project Associate) mini-grants. This program proved that small amounts of money can be used to implement change, if faculty are motivated and have local financial and administrative support. As was described in this study, these grants of $5,000 per course, matched by $5,000 from the institution, provided a small but very significant incentive, including both rewards and recognition for those desiring to implement PLTL. Fifty-five of 81 WPA grant recipients reported that they would not have implemented the workshops without the grant. Nevertheless, 80 of the 81 respondents said they intended to continue the program after the grant. These numbers indicate that the while funds from the PLTL project were important to faculty trying the method, the change remained in place after the external funds were depleted.

The process of preparing proposals and reporting back to the project created a mentoring structure to assist with implementation and fostered an additional level of commitment from the implementers. Requiring written proposals became an important mechanism for providing preventive intervention, before funding was made available, because the proposals revealed flaws in implementation plans. Connection to a national network of like-minded faculty provided a support structure for faculty while they were in the process of first implementing PLTL in their own courses.

Grantees knew that they were expected to use the project resources responsibly and in almost all cases they did so. Through the mini-grants, PLTL spread successfully through two-year and four-year private and public institutions across the country. Dissemination efforts were multiplied as WPA recipients made their experiences public through presentations and publications (Baez-Galib et al., 2005;

Stewart et al., 2007). Projects like the Multi-Initiative Dissemination Project (MID) of which Varma-Nelson was a senior partner, and Project Kaleidoscope (PKAL) proved to be ideal places for new implementers to present workshops and share their experiences and results with other interested faculty.

9.2.1 Disciplines

PLTL has made progress in gaining adopters beyond chemistry. The first NSF National Dissemination grant funded in 2000 included faculty from biology and physics as well as chemistry as co-principal investigators. Since then, the method has been successfully implemented in biology, mathematics, computer science, and other disciplines. The PLTL founding group placed a high priority on cultivating new leaders for the different disciplines.

There have, however, been issues in gaining momentum beyond chemistry. There are several reasons for this. At the introductory levels, the disciplines are still quite distinct. Chemists attend conferences sponsored by their organizations, as do biologists, mathematicians, and others. New ideas about pedagogy are generally transmitted through these meetings and the associated publications. Chemistry faculty have had considerable success promoting PLTL. But since the opinion leaders with real influence are generally found within the discipline, it is difficult for faculty from one discipline to present ideas to those from another. A careful study of dissemination strategies (Foertsch et al., 1997)) found that faculty members are particularly influenced by those in their field with scholarly reputations and whom they view as colleagues. So it is now up to the adopters in other disciplines to create a PLTL presence within their professional organizations. This has begun to take place. The computer science network funded by NSF, described in Chapter 7, has used PLTL as a key strategy in its efforts to attract women to computer science.

Larson and Meyer (2007) provide data, based on survey research, indicating that college professors recognize opinion leaders within their departments and in fact listen to them about pedagogical initiatives. The PLTL project did not deliberately seek out opinion leaders in the dissemination process, but we now view this as useful strategy and recommend it for further dissemination of the model.

A second major issue in advancing PLTL in disciplines beyond chemistry has been the development of materials for workshop learning. The development and publication of materials for general and organic chemistry were important factors in the successful dissemination of PLTL in chemistry even though most of those reporting said they modified the materials (54 of 76 respondents used the project-developed materials at least in part.)

The spread of the method was slow in biology and mathematics because preparation of materials proceeded more slowly in these areas. In biology the slow development was probably due in part to the fact that there is a wide range of possible topics in introductory courses and it was not easy to provide even a sample or outline that would be generally useful. In mathematics, the challenge is the variety of foundation courses. In both these disciplines, materials had to be developed or adapted from textbook materials at the local level, and this slowed the pace of dissemination. Mathematics materials are now available on the PLTL website.

We are not ready to present a set of critical components for dissemination, as we do for implementation and institutionalization. There is more to be done in developing a theory of diffusion of educational practices. We can, however, conclude this discussion with the lessons learned from the dissemination of PLTL.

- The four stages used by the project provided a useful framework, regarding both the theory and practice of dissemination.
- Mini-grants were valuable in the dissemination of PLTL and we believe they would be useful in other initiatives.
- Local leadership and adaptation helps increase commitment and adds to the vitality of implementation.
- Each discipline has needs of its own and requires its own leadership.

9.3 Issues Regarding Institutionalization

Based on the analysis of PLTL data we propose the following as critical components for institutionalization.

1. Implementation according to the model and on-going fidelity to the model
2. Administrative support and funding
3. Demonstrated success
4. Fit with the institution's mission and practice
5. A core group of committed faculty in a department or an institution

We consider these to be necessary conditions for the survival of PLTL, but cannot say whether they are sufficient. Further studies may uncover other factors that we have overlooked. There are issues associated with each of the factors we have mentioned, and we will briefly review them.

The fact that PLTL developed a clear model made it transportable. But Rogers (2003) and others have pointed out that implementers need ownership and this often means adaptation or re-invention. Developing new materials and introducing workshops into new courses provided opportunities for creativity and ownership. As noted in several chapters, some adaptations work better than others. When the method is used in biology, workshops tend to focus less on problem-solving and more on a review of conceptual material. In some advanced courses the method has been used in conjunction with the case study approach and problem-based learning. These and other adaptations have worked well. But in other cases, when the method has been adapted as a drop-in review session the original method has been diluted to the point where it is not recognizable and is not effective. PLTL also does not work well when used to fill a remedial need because the program's strength derives from the fact that students at various levels assist one another based on social and cognitive development. In addition, for minority students negative stereotyping often undermines remedial activities (Bonner & Bailey, 2006).

Ongoing funding is essential to maintain PLTL. When this funding must be requested and reapproved each year, those directing the program are likely to lose their motivation, and the project may fail even if it enhances student success. The effect of reliable funding is also important because it demonstrates a level of administrative confidence that helps weave the program into the fabric of the institutional culture.

Demonstrated success of the program is essential for institutionalization. This success means not only that the program has been implemented without major problems. It is also essential that the input of time, energy, money, and reputation are viewed as more than compensated by the output of academic performance, motivation, teamwork, and the many benefits to the peer leaders that persist after they have graduated and entered the professional world.

Fit with the institution's mission may sound vague, but it is real. While all colleges and universities are dedicated to teaching and learning, not all include an articulated vision, structures and resources that support new ventures. Commitment to teaching and learning may be translated to practice in different ways at different kinds of institutions—community college, private college, public institution, etc.—and at particular institutions with their own histories and priorities. We have described many institutions for which PLTL was a good fit because faculty felt responsible to help each student learn. By contrast, in some colleges and research universities the prevailing climate dictates that it is best if the hundreds and perhaps thousands of students in the gatekeeper courses sink or swim on their own. "Babying" them will only create problems in the advanced courses. PLTL will not thrive in an institution where this is the prevalent mentality.

Our final essential ingredient is a core group of faculty committed to learn from each other's experiences and to form a support structure. Continuity, mutual support and pursuit of a common goal are important. A group is also essential because of the mobility and freedom of college professors. Young professors often move several times before settling down in a career at one institution. Even those who are senior and appear settled sometimes move. When a faculty member retires or leaves an institution, no one can insure that the new person will maintain Peer-Led Team Learning. If the project has a strong foundation and is viewed as important within the department, then it is more likely that a new person will buy into the project.

9.4 Summary

We conclude with notes on two areas of the project describing what we consider to be the major benefit derived from PLTL, and the most significant threat to the continuance and institutionalization of the program.

PLTL has many strengths. The program improves retention and academic performance, increases student interest in the coursework, fosters various interpersonal skills, develops leaders, and acts as a catalyst in intellectual development and academic involvement. Overall the outstanding benefit of PLTL is increased student involvement and engagement. We are convinced by the analysis and arguments of Astin (1993) and others that involvement is in fact the single most important factor accounting for a young person's success in college.

We believe that the major downside or threat to PLTL is the fact that faculty members are not connected to one another in their teaching. We have seen that the independence and self-directedness that are so prized in the academic profession can inhibit programmatic progress. It is difficult for an institution or department to promise that PLTL or any other program that moves away from traditional entrenched pedagogy such as the lecture, will survive. PLTL like many other programs may or may not survive at an institution depending on the will of the professor, and more importantly on the interests of his or her successor. Paying attention to the lessons learned from more than a decade of experimentation with PLTL can improve the chances of survival.

In conclusion the work of the national PLTL project provides models for local implementation, dissemination and continuing sustainability. We recommend the following to an institution interested in planning, implementing, and sustaining Peer-Led Team Learning:

- Offer an in-house program of mini-grants for faculty.
- Implement a peer-leader training course that deals with pedagogy and content with significant faculty leadership. Make this a credit-bearing course.
- Build a team of PLTL practitioners who work together to initiate new faculty to the program, and to institutionalize of the program.
- Do comparative grade and retention based studies, and use recognized assessment instruments such as those provided by PLTL, SALG, or others that are being developed and published.
- Include a partnership with students in planning and implementing the project.
- List PLTL workshops on the departmental course schedule.
- Acquire space that is suitable and friendly for the workshops.
- Create a center for workshop-related education with clerical and pedagogical support.
- Recognize and reward the efforts of faculty and peer leaders for their contributions to improved teaching and learning.

Chapter 10
Evaluation Strategies

The evaluation of Peer-Led Team Learning has been discussed throughout this report. In this chapter we summarize the evaluation plans, methods, and outcomes as they were conducted in conjunction with the development of PLTL, looking particularly at lessons relevant to other projects. We talk about "evaluations" in the plural since there were so many people and institutions involved and because so many different aspects of the program were evaluated. We also present a number of evaluation instruments that may be adapted to other programs. Finally, we demonstrate how the evaluation contributed to the development of a model that could be used to assess implementation and support dissemination.

The evaluations covered overlapping areas but they are considered separately for purposes of analysis. These areas occurred in a cyclic rather than sequential manner. Dissemination, for example, began even before the first implementation grant, as the originators of the project invited colleagues to join them, and program implementation was frequently revisited as the program expanded within an institution, migrated to new sites or was adapted to new disciplines and circumstances. The evaluation areas, most of which can be adapted to other initiatives, are as follows:

- Program implementation
- Student and student leader experiences
- Student performance
- Program monitoring
- Faculty response
- Administrative response
- Impact on peer leaders
- Dissemination
- Institutionalization

10.1 Program Implementation

When faculty members at the first institutions included workshops as part of their courses, they found that the process and the requirements were more complicated than they had anticipated. Gafney began listing questions that reflected faculty concerns. Discussions made it clear that these questions fell into a number of categories. In June, 1996 a list of nine essential elements was drafted and sent to PLTL faculty with 37 implementation questions for their consideration.

Not all of these issues were analyzed in detail. But they were at least recognized, and they became an important tool for the external evaluator. Some were discussed more than others, and within a year the nine "essential elements" had been reduced to six "essential components," which later became the critical components we have discussed throughout this study. They were presented in August, 1997 with questions for further review.

Some of the questions led the participants to consider important features that should belong to workshop courses, for example, constructing the syllabus so that workshops and lectures are synchronized. Some of the questions were included to prompt ideas and consideration of what might be done, for example, using part of the lecture for small group work. Still other items look for information useful in the aggregate, for example, how much additional time do workshop activities require of the lecturer.

The complete list of questions submitted to the implementers at that time is as follows.

Workshop chemistry: Questions to add detail to the essential components.

A. *The workshop is integral to the course, coordinated with the other elements.*

1. Does the workshop cover the lecture material for that week?
2. Are workshops sometimes used to remediate or cover prerequisite skills?
3. Do you sometimes use part of the lecture for small-group work?
4. Has lecture time been changed with the introduction of workshops?
5. Are students required to attend the workshops?

B. *The workshop materials are challenging and integrated with the other course components, intended to encourage active learning and to work with groups.*

1. How were the workshop materials designed and prepared?
2. How are they related to the lecture and to the textbook?
3. How are they used?
4. How are they evaluated?
5. Are workshop problems like the problems found on tests?
6. Do workshop materials include challenging non-routine problems?

C. *The workshop leaders are well trained and supervised, with attention to knowledge of chemistry and teaching/learning techniques.*

1. What training is provided in knowledge of chemistry needed?
2. What training is provided in collaborative teaching/learning skills?
3. How is supervision of the workshop leaders done?
4. Is problem solving emphasized?
5. Are activities planned for exploratory discussions?
6. Do students work with models?
7. What team/collaborative learning activities are used?

8. How is workshop time divided among: presentation by the leader; paired or group problem solving; individual work; student presentations; tutoring; computer/lab activities; other?
9. What methods are used to get students working together?
10. How are students assigned to groups?
11. How are less able students accommodated?
12. How are more able students accommodated?

D. The faculty teaching the course are closely involved with the workshops and workshop leaders.

1. How much extra time do workshop preparation and activities take?
2. What other faculty and professional resources contributed to the workshops?
3. For what amount of time does the professor meet with workshop leaders?
4. Is someone else delegated to train and supervise the workshop leaders?
5. Is there a support course in learning and group processes for workshop leaders?
6. Has the workshop approach led to revisions in the assessment of student performance?
7. What administrative obstacles might delay the adoption of the workshop model as part of the course of studies?
8. Are courses with workshops increasing?
9. How do other faculty members view the workshop approach?

E. The organizational arrangements including the size of the group, space, time, noise level, etc. are such as to promote learning.

1. How often are workshops held?
2. Is the space adequate?
3. Is noise a problem?
4. How long are workshops?
5. What is the best size for cooperative groups?
6. On average, how many students are in a workshop group?

F. Administrative personnel and structures support the workshop.

1. Are teaching and curriculum work valued for promotion?
2. Are adjustments made to accommodate workshop courses in the catalog?
3. Is on-going funding available for workshop leaders?
4. Are space and time adjustments made to suit workshop courses?
5. Are courses in pedagogy included for student leaders?

As has been mentioned throughout this study, the identification of the six critical components, through the evaluation, was viewed as an important step by those implementing and promoting PLTL. These components, and the details associated with them, provided a model used first as a roadmap for implementers, second to describe the method to those who were interested, and third to review implementation after adoption. In Chapter 2 we recounted specific situations illustrating how failure in some aspect of the critical components led to radically diminished success, and a lack of enthusiasm on the part of students and faculty.

Development and use of the critical components is clearer in retrospect than it was in practice. But the process can be used as a model by others as they develop, implement, and disseminate educational programs or new approaches to teaching and learning. The following recommendations are extracted from the process used in the development of the critical components for PLTL:

1. Have clarity about the overall goals. For PLTL, the goal was to improve academic performance based on traditional norms and to promote more involved, active learning in which students worked productively with one another.
2. Implement at a number of sites, asking questions and comparing experiences. Make a list that includes those features, activities, or parts of the program that are very important. For PLTL, the initial implementers met regularly and the external evaluator collected information and developed questionnaires based on the information gathered. The critical components for most projects will fall into categories such as: organizational arrangements of time, space, and special materials; student activities; instructor activities; training and supervision of project personnel or students and technological needs. These are not exhaustive but should provide a good starting point as a project seeks to identify its own essential features.
3. Collect the activities and program features into appropriate categories. Review and revise the categories until they are of a manageable format and number to study and use, probably between three and eight. For PLTL the six critical components have endured for 10 years.
4. Review the results with all participants and collect data on all aspects of the project until agreement is reached that these are in fact the components required for success. Again, for PLTL the structure of implementation and the variety of sites provided ample opportunity to review implementation issues.
5. Test the critical components by observing whether they are in use. If they are not, determine the impact on the project. It may be found that a particular component is not essential; or it may be that the diminished success of the project is readily seen if a particular component is not in use. For PLTL, site visits by Gafney uncovered convincing evidence that the failure of any critical component would jeopardize the project.

Careful examination of implementation goes back to a study performed over 30 years ago (Gross et al., 1971). This project reviewed a number of impact studies in which outcome measures determined that educational interventions had not had any significant outcomes. The studies were based on before and after measures, comparisons with control groups, or achievement measured against goals. What the studies failed to consider, and what was uncovered by the more careful program documentation, was that the interventions had in fact never been implemented to a degree that might have produced the desired results. For example, prescribed educational materials were never used, teacher aids were not trained according to plan, or prescribed teacher behaviors were not initiated. The conclusions should have been not that a particular initiative failed, but that it had not been introduced.

More recently a related point was made by Elmore (1996). He found evidence that in the United States new approaches to teaching and learning are rarely implemented in anything but a small fraction of schools and classrooms. He calls the situation a problem of scale and believes the reasons are embedded in the teaching/learning environment. He believes that the solution to this problem is to create structures and incentives to promote faculty development and provide appropriate support in implementation.

There is reason to believe that the situation is even more prevalent at the college level, that is, it is more difficult to change a college professor's approach to teaching than that of a school teacher. This is true because institutions of higher education do not generally require regular participation in professional development to improve teaching and learning. For individuals or groups interested in change, however, the PLTL model does present a clear-cut model for implementing programs and monitoring their progress.

10.2 Student Experiences with PLTL

The first evaluations of PLTL were through student focus groups. We wanted to find out whether the workshops made a difference in the way students learned and the way they viewed themselves as learners. These focus groups were the first evaluation activities. They provided insights and data used to design surveys and interview protocols.

The following list shows the questions used to promote discussion in early focus groups about the workshops:

1. Which workshops did you find most helpful? most interesting? Why?
2. Which workshops were most difficult? Why?
3. How did the workshops improve your understanding?
4. How were the workshops related to the class lectures? Did the professor refer to the workshops? Did the workshop leader refer to the lecture material? the text?
5. What kinds of cooperative activities did you do in the workshops? How were the cooperative activities helpful?
6. Did you become involved with a study group as a result of the workshop?
7. Did the workshop approach get you to participate more than you do in class? How?
8. How did the leader conduct the workshop?
9. What might the leader have done differently?
10. Will you try to take other courses that have workshops?

The following is a report written in 1995 based on the initial focus groups. The experiences illustrate the benefits to these students and student leaders. It was important that their comments then be refined and structured into survey format so that the impact of the workshops could then be evaluated statistically with much larger populations.

10.3 Focus Groups Findings

The focus groups were held in the spring of 1995 to gather information about students' and leaders' reactions to the workshops that are part of General Chemistry 103 at the City College of New York. There were three focus groups, each with nine or ten participants. Two of the groups were of leaders; one was of students. Questions developed and modified in consultation with the professors were used to promote discussion.

10.3.1 Overview: Lecture and Workshop

The use of workshops to complement lectures was enthusiastically endorsed. Leaders and students repeatedly said that many times they could not understand material as presented in the lectures. They thought that the professors sometimes assumed knowledge, "started in the middle," or "took short cuts." They wished that the professors would "stay on a topic a bit longer." They gave reasons why they are reluctant to ask questions: the anxiety associated with speaking in a large group; a fear that the professor and their peers would think that a question was "stupid;" a feeling that the professor will simply explain the material again in the same way; an inability to formulate a question; lack of time; inability to get the professor's attention.

In contrast to the above, the students found that in the workshops anxiety is reduced, leaders are accessible, and peers are supportive. They no longer feel isolated in the learning enterprise. Their incomplete knowledge is no longer felt as a liability, but is actually an asset because it permits them to join in group activities, helping and being helped according to need. Some of the leaders contrasted this approach to other courses in which "students might not say anything for a whole semester."

10.3.2 Workshop Methods and Dynamics

The leaders and students agreed that one-on-one help with problem solving and with new concepts was a great benefit. The workshop leader is viewed as a peer, sometimes a friend. It was said frequently that the leader was able to explain things "in a different way." This different way meant several things: supplying background information; breaking the material into smaller chunks and showing connections; using different vocabulary and examples. Several of the leaders found it rewarding to help students strengthen their math skills.

The leaders were successful because learning was individualized with questions possible at each step and because the affective environment was so different from the classroom. The students like the leaders because they are close in age, "know where you are coming from," and "the way you understand things." They also like the informal workshop setting, and the atmosphere that encourages them to speak their minds. Both leaders and students believed that since the leaders had recently completed the course they were familiar with the material and could explain it well.

There was agreement in all groups that students started out feeling and acting alone, carrying over their classroom attitudes to the workshops. But within a few weeks behaviors changed. Workshop leaders asked individuals to explain problems. Students became increasingly confident. They began questioning and helping one another. They found it beneficial that sometimes the same idea would be expressed in different ways by different students. They also noted the way in which learning is deepened through discussion: "If you can explain it, you understand it."

One workshop leader noted that cooperative activities are common in research and business, so it is important that they and the students learn to work together to solve problems and to build on each other's insights.

The focus groups were followed by the use of surveys to look at the experiences of students and leaders statistically and across a number of sites. The survey form for students as presented here is one that evolved from the one shown in Chapter 2. We wanted to gather some very practical information about how students experienced the workshops and whether these experiences were consistent with the workshop goals. Some of the important issues were:

- Whether students believed that the workshops helped them learn and improved their grades.
- Whether from the students perspective the critical components were operative, for example: useful materials, connections to the lecture, adequate organizational arrangements, etc.
- Whether the workshop dynamics were appropriate, with students actually working together in productive ways under the guidance of the leader.

In using the leader survey, we wanted to compare experiences of students with workshop leaders, but we also wanted information about the group size, time, and a number of other particulars about which the leader would have special information and insights.

These surveys were made available and used by individual sites to assess their PLTL activities, and sometimes to make comparisons across sites or over time.

Student Survey

Peer Led Team Learning **Institution** ____________________

Course Name ____________________ **Professor** ____________________

For each item, circle the number that corresponds to your response:
5 = strongly agree; 4 = agree; 3 = neutral (no opinion); 2 = disagree; 1 = strongly disagree.

	Agree				Disagree
1. The workshops are closely related to the material taught in the lectures.	5	4	3	2	1
2. Workshops help me do better in tests.	5	4	3	2	1
3. Interacting with the workshop leader increases my understanding.	5	4	3	2	1
4. The workshop materials are helpful preparation for exams.	5	4	3	2	1
5. Workshop materials are more challenging than most textbook problems.	5	4	3	2	1
6. I believe that the workshops are improving my grade.	5	4	3	2	1
7. I regularly explain problems to other students in the workshops.	5	4	3	2	1
8. Interacting with the other group members increases my understanding	5	4	3	2	1
9. I would recommend workshop courses to other students.	5	4	3	2	1
10. In the workshops I am comfortable asking questions when I do not understand something.	5	4	3	2	1
11. The lecturer encourages us to participate in the workshops.	5	4	3	2	1
12. The workshops are often dominated by one or two students.	5	4	3	2	1
13. Noise or other distractions make it difficult to benefit from the workshops.	5	4	3	2	1
14. Students who are uninterested or unmotivated make it difficult for others to benefit from the workshops.	5	4	3	2	1
15. I felt comfortable with the workshop leader.	5	4	3	2	1
16. The workshop leader is well prepared.	5	4	3	2	1
17. I am uncomfortable asking questions in the lecture.	5	4	3	2	1
18. The workshops are a big help in solving problems.	5	4	3	2	1
19. I would like to be a workshop leader in the future.	5	4	3	2	1
20. In the workshops I enjoyed interacting with the other students.	5	4	3	2	1
21. The workshop experience led me to join formal or informal study groups related to other courses.	5	4	3	2	1

22. On average, I spend the following number of hours per week studying (in addition to time spent at lectures and workshops):

(1) 0–2 hours (2) 2–4 hours (3) 4–6 hours (4) 6–8 hours (5) 8–10 hours.

These items are about the materials used in the workshops. Use the following scale:

5 = materials are excellent for meeting this objective; 4 = materials meet this objective very well; 3 = materials meet this objective rather well; 2 = materials somewhat meet this objective; 1 = materials do not meet this objective at all.

The materials are:

23. Well connected with the lecture	5 4 3 2 1
24. Challenging	5 4 3 2 1
25. Developed to review fundamentals	5 4 3 2 1
26. Useful for group work	5 4 3 2 1
27. Motivational	5 4 3 2 1
28. Helpful for individual study	5 4 3 2 1
29. Useful for reinforcing concepts	5 4 3 2 1

Rate each of the following activities according to the amount of workshop time devoted to it. Use the following scale: 5 = most of the time; 4 = a large amount of time; 3 = a moderate amount of time; 2 = a small amount of time; 1 = almost no time.

30. The workshop leader presents ideas and methods.	5 4 3 2 1
31. The leader responds to student questions.	5 4 3 2 1
32. Students work on problems in pairs or small groups.	5 4 3 2 1
33. Students work on problems alone.	5 4 3 2 1
34. Students present solutions.	5 4 3 2 1
35. Hands-on activities.	5 4 3 2 1
36. Technology and computer simulations.	5 4 3 2 1

Thank you for your participation.

Leader Survey
Peer Led Team Learning **Institution** ____________________
Course name ____________________ **Professor** ____________________

1. How often do workshops meet? ____________________
2. What is the scheduled length of a workshop meeting? ____________________
3. For how long do you usually meet? ____________________
4. How many students are enrolled in your workshop? ____________________
5. On average, how many students usually attend a workshop? ______________
6. What do you think is the best number of students for a workshop? ____________
7. Attendance at workshops (is, is not) a course requirement.
8. About how much of your time per week is taken by workshop preparation and activities, not including the workshop itself? ____________________
9. Please describe the activities as they take place in a typical workshop?

Rate each of the following activities according to workshop time devoted to it. Use the following scale: 5 = most of the time; 4 = a large amount of time; 3 = a moderate amount of time; 2 = a small amount of time; 1 = almost no time.

10. The workshop leader presents ideas and methods.	5 4 3 2 1
11. The leader responds to student questions.	5 4 3 2 1
12. Students work on problems in pairs or small groups.	5 4 3 2 1
13. Students work on problems alone.	5 4 3 2 1
14. Students present solutions.	5 4 3 2 1

15. Hands-on activities such as use of models 5 4 3 2 1
16. Use of technology or computer simulations 5 4 3 2 1
17. Are workshop problems good preparation for tests? Please describe.
18. Do workshop materials include challenging problems? Please describe.
19. Were the workshop materials too difficult or too easy for students in your group? If so, what did you do?

The next items refer to the materials used in workshops. Circle a number from 1 to 5 according to how well they meet each objective: 5 = materials are excellent meeting this objective; 4 = materials meet this objective very well; 3 = materials meet this objective rather well; 2 = materials somewhat meet this objective; 1 = materials do not meet this objective at all.
The materials are:

20. Well connected to the lecture 5 4 3 2 1
21. Challenging 5 4 3 2 1
22. Developed to review fundamentals 5 4 3 2 1
23. Useful for group work 5 4 3 2 1
24. Motivational 5 4 3 2 1
25. Helpful for individual study 5 4 3 2 1
26. Useful for reinforcing concepts 5 4 3 2 1
27. What methods are used to get students working together?
28. What did you do for students having difficulty?
29. Did students sometimes discuss personal problems with you? If so, how did you respond to them?
30. What training and support are provided to leaders in how to run workshops, for example in group dynamics or instructional processes?
31. What training and support are provided to the workshop leaders in the knowledge of the discipline?
32. What training and support are provided to the workshop leaders in theories of learning and related methods of teaching?
33. What parts of student leader training have been most useful? What do you need more of?
34. How do you interact with the professor teaching the workshop course?

The written surveys of students and student leaders did in fact yield valuable information, not only about student satisfaction with the program, but also about the critical components. These surveys in conjunction with faculty interviews and observations offered multiple perspectives and the opportunity to gain a more comprehensive view of how well the program was progressing.

We recommend this approach for any curriculum or teaching/learning innovation. The best way to find out what students are experiencing is to ask them. This information is easy to obtain in the survey format, or in an approach combining surveys, interviews, and focus groups.

10.4 Student Academic Performance

There were many ways to assess the impact of PLTL on student learning. These are described with examples in Chapter 2. We will now summarize the different approaches to gathering data about student performance in the form of recommendations, listing some of the advantages and disadvantages of each:

1. Comparisons with historic grade data. Assuming that tests, student academic levels, and grade-related course materials and activities have been relatively constant, compare grades for students enrolled in workshops with those of previous years when the workshops were not part of the course. Statistically, this method may be the simplest and easiest to use.
2. Randomly assigned groups. Divide the class, randomly assigning one group to workshops and a matched group not to participate in workshops. Except for workshops (or other intervention), treat the two groups equally. This method has statistical power and may be used to compare groups with and without workshops or to compare workshops with recitations or some other intervention. But students may not be receptive to the assignment of two extra workshop hours per week.
3. Compare self-selected participation in the workshop groups with the rest of the class. This is often done in a pilot program for which resources do not permit implementation with an entire class. In fact, a number of successful PLTL program have been organized on a self-selected basis. Some argue that the self-selecting participants are likely to be more motivated. This could be true but as the Northwestern study showed, these students may also be more anxious about their academic progress.
4. Do a correlation of attendance at workshops with grades. This has the advantage of looking not only at enrollment in the program but also at participation on a weekly basis. As noted in Chapter 2, PLTL attendance at workshops was highly correlated with grades.
5. Compare scores on the standardized tests. Such tests are available, for example, from the American Chemical Society, Division of Chemical Education Examinations Institute, http://www4.uwm.edu/chemexams/. Scores can be compared for those using PLTL and a control group—either historical or contemporary. Results of this method are presented in Chapter 2. Standardized tests have the advantage of being well known across the country. The disadvantage is that they may not provide a close match with the objectives of a particular course, and may not be available for some courses.

It is important in any system of evaluation to control as much as possible for differences in instructors, material presented, and grade-related data such as extra credit for workshop attendance. But the problem with introducing too many controls, as mentioned in Chapter 2, is that the situation as created may become artificial rather than "real world." As Chen (2005) points out, a real-world evaluation may be preferable to one that is highly controlled precisely because it exists in a real teaching/learning setting rather than a learning laboratory.

10.5 Program Monitoring and Small Grants

Program monitoring is described by Chen (2005) as the periodic collection of quantitative data requested by a program evaluator and carried out by the implementers. He also discusses process monitoring which includes the experiences of participants and outcome monitoring in which goal-related performance data are collected.

Part of the program monitoring of PLTL, as described above, included the collection of student and leader responses to the program and comparative data on grades and other academic outcomes. There was additional program monitoring in connection with the WPA mini-grants used in dissemination. The program evaluator, Gafney, worked closely with Varma-Nelson, the principal investigator who supervised the mini-grant program. Thus, assessment and data collection were built into the structure of the grant process and program implementation, from the outset. These mini-grants and the variety of strategies employed provided a "laboratory" in which to study implementation and dissemination. Guidelines for the submission of proposals were as follows.

10.5.1 WPA Request for Proposals

The Workshop Project Associate (WPA) program provides funds to assist faculty and learning specialists to develop and implement a peer-led team learning course at their institution.

Proposals may be submitted for support of peer-led team learning course development in any field of science and in mathematics.

Proposals are invited from organizations in the United States and its territories: two-year colleges, four-year colleges, and universities. Applicants must show evidence of familiarity with the peer-led team learning model of instruction. This can be demonstrated through attendance at a PLTL short course or workshop. Evidence of a mentoring relationship with an experienced peer-led team learning instructor is desirable.

10.5.2 Proposal Preparation

The written proposal should contain the following information, assembled in the order indicated:

1. Cover sheet

A one-page cover sheet indicating the following:

A. Name and postal address of the principal investigator (PI)
B. Telephone number, fax number, and e-mail address of PI

C. Name and postal address of the organization to which the award should be made
D. Discipline under which the proposal will be evaluated
E. Title of project
F. Requested amount

2. Project description

Text in this section of the proposal should be double-spaced. Use standard margins, 12-point font, and print only on one side of the page. Limit the project description to *no more than five pages*. The description of the project should contain an explicit statement of plans for meeting the critical components of peer-led team learning, including:

A. Description of how the peer-led team learning component of the course will be integrated with other course components
B. Description of how the course instructor will be involved with the peer-led team learning component of the course
C. Plans for recruitment and training of leaders
D. Plans for choice of materials to be used in the course
E. Plans for obtaining appropriate organizational arrangements (time, space, group size, etc.)
F. Evidence of institutional support (such as matching funds for leaders, release time for faculty, recognition for teaching, etc.)
G. (Optional) Describe plans to develop and test Workshop Modules (see "F" under Budget and Budget Justification)
H. The philosophy of teaching guiding the proposed implementation
I. Description of experience with collaborative/cooperative learning or other curricular revision projects
J. Description of the course in which the peer-led team learning model will be adapted
K. Plans for on-going evaluation (evaluation should utilize standard Project materials (see website), although more extensive plans are welcomed);
L. Description of contribution from the institution's learning center (if any);
M. Plans for dissemination of the PLTL implementation.

3. Biographical sketch

Provide a biographical sketch of no more than two pages for the PI. Include items relevant to experience with curricula similar to peer-led team learning and/or curriculum reform.

4. Budget and budget justification

Provide a one-page budget and a one-page budget justification on separate pages. Funding up to $5,000 per person (proposal) and $10,000 maximum for a discipline (department) at an institution will be considered. Requests for Workshop leader salaries must be matched on a one-to-one basis from the institution. Show matching institutional funds explicitly on your budget page. *No indirect costs should be included.*

The PLTL Project will provide evaluation assistance by the Project Evaluator. Acceptable budget categories include: (a) Workshop leader salaries, (b) support for preparation of materials, (c) travel for dissemination, including the presentation of posters, papers, workshops, or short courses, (d) evaluation, and (e) support for participation of a learning specialist, (f) an additional $1,000 stipend for faculty to develop at least one original workshop module and $200 stipend for students to test the new module (biology courses only).

5. Statement of tasks to be performed with time lines

Provide a statement of tasks to be performed, i.e., training of peer leaders, materials development, evaluation, dissemination, etc. All activities must have specific time lines.

10.5.3 Proposal Review Criteria

The general review criteria are: how well the proposal meets the critical components of peer-led team learning, evidence of understanding the peer-led team learning model, quality of plans, probability of successful implementation, and the other items requested in the project description section. The panel will also consider the qualifications of the personnel submitting the proposal.

Reviews are conducted by a panel appointed by the WPA officer for each discipline. Each panelist writes an individual review for all proposals assigned to the panel. The reviews are used by the WPA officers to make final funding decisions.

Proposals for the WPA mini-grants were reviewed by three experienced members of the PLTL team, the reviews were made available to applicants. The criteria for reviewing proposals are contained in the following document.

Reviewer's Name:
Reviewer #
Title of Project:
Principal Investigator:
Requested Amount:
Proposal Review Cycle ending with Submission Deadline:
Please address the following as you review this proposal:

A. Plans for choice of materials to be used in the course
B. Plans for recruitment and training of leaders
C. Description of how course instructor will be involved with the peer-led team learning component of the course
D. Description of how the peer-led team learning component of the course will be integrated with other course components
E. Evidence of institutional support (such as matching funds for leaders, release time for faculty, recognition for teaching, etc.)
F. Plans for obtaining appropriate organizational arrangements (time, space, group size, etc.)

G. The course in which the peer-led team learning model will be adapted
H. Appropriateness of plans for on-going evaluation (evaluation should utilize evaluation materials already developed by the project, although more extensive plans are welcomed)
I. The philosophy of teaching guiding your implementation
J. Description of experience of collaborative/cooperative learning or other curricular revision projects
K. Plans for dissemination of your implementation
L. Description of contribution from your institution's learning center (if any)

Qualifications of Personnel:
Other Comments:
Recommendation for funding:

--- Strongly Recommend
--- Recommend
--- Recommend with the following reservations
--- Do Not Recommend funding at this time

Each of the more than 90 individual faculty who were awarded these WPA (Workshop Project Associate) grants was asked to complete the following report form.

WPA Report

Today's date: _____

Principal Investigator: ____________________
Institution: ____________________
Grant Period: ____________________
Email: ____________________
Co-PI's (if applicable): ____________________
Email (Co-PI): ____________________
Organizational Arrangements

Please answer the following for each PLTL (Peer-Led Team Learning) course.

Course name	Month/year completed	Instructor name	Institution	Number of students	Number of peer leaders	Learning specialist name

Average size of workshop groups:

Materials
What materials did you use or develop?
If adapted, what were the major changes you made in materials?

Training and Supervision of Workshop Leaders
What recruitment and initial training did you do with peer leaders?

What type of on-going training and supervision did you do?
Was anyone else involved in the training? If yes, what role did they play?

Institutionalization and Dissemination
What are the plans for the continuation of PLTL on your campus?
Have other faculty members within or outside your department expressed any interest in PLTL? If yes, please describe.
Have you disseminated the results of your project in any way?

Problems
Did the project involve what you had expected or more work than you expected?
What difficulties or problems were encountered?
Who helped you when you had problems with PLTL?
What changes did you make regarding your original proposal?

General
Do you judge the PLTL program to have been a success at your campus?
What did you find to be the principal strengths and weaknesses of the model?

WPA Procedures and Future Grants
Would you have implemented the PLTL methodology in the absence of the WPA program?
Should the WPA program give more grants with smaller awards, fewer grants with larger awards, or is the funding level reasonable for the goals of the program? Explain.
How can the program be improved to better achieve the goal of dissemination?

Have you received additional funds for PLTL? If so, where did the money come from. (i.e. your institution, other funding organizations)?

The information gathered through this form was useful to the project. As noted in the Chapter 4, data were collected and analyzed for a variety of areas—recruiting and training leaders, organizing the workshops, developing materials, student outcomes, and faculty involvement. Just as important as these data were the descriptions of benefits, problems, modes of adapting the workshops, and local dissemination. This new information provided fresh perspectives about the program and in addition, as Chen (2005) notes, "the monitoring requirement increased the level of commitment by participants."

The WPA reports served another very important purpose: they uncovered situations for further study through the use of follow-up phone interviews and site visits. In the reports, grantees presented particular issues related to local departments, disciplines, type of institution, or unusual approaches to the workshops. The evaluator could then follow up, looking at the details to complete a picture at the local level and also add to the overall project data.

Based on this project, we would recommend program monitoring that collects useful information and in which benefits to the participants are apparent. With regard to PLTL, participants generally wanted to use the student and leader questionnaires, and many wanted to do comparative performance studies. In some cases, several requests were needed to obtain completed WPA forms. But participants were readily available for phone interviews and site visits, and in some cases networks were formed based on the information gained and processed by the project administrators and evaluator.

10.6 Faculty Response to the Program

Since Peer-Led Team Learning calls for a number of changes in faculty attitudes and behaviors, we wanted information about their experiences with the workshops. It was clear that, in general, instructors had positive feelings about the program, but it was important to gather as much detail as possible. It was evident, based on initiatives such as the Keller project, that even worthwhile programs are sometimes abandoned because they require greater time, effort, or role changes than many faculty members will accept. Several methods were used to gather information about faculty experiences with the program, including:

- Visits to selected sites with interviews and group discussions
- The WPA reports, described above
- Phone interviews
- An email survey on institutionalization
- On-line surveys

Several on-line survey services were used and they proved to be a good alternative to paper and pencil questionnaires for faculty who are generally in front of the computer and find this approach easier and quicker than paper forms that require more handling and mailing. The raw data from an item asking about how PLTL was adapted or related to other initiatives is shown below and illustrates the data-collecting power of the method. Sometimes an open-ended question, with no preconceived ideas about answers, yields a wealth of data with rich insights into the program.

If PLTL has been adapted to accommodate other teaching/learning initiatives, please describe briefly.

- PLTL was used in an advanced chemistry class, within the lecture.
- PLTL has been adapted for shorter sessions with guided inquiry materials that can accommodate the maximum number of students per peer leader via cooperative learning.
- PLTL is being used in our large, sophomore laboratory to analyze data from a challenging experiment.
- Tried a 'POGIL' approach in GOB course in fall. In retrospect, it was not a real improvement over the PLTL of previous years.
- We have tried to include instruction in Technology Literacy and Information Literacy as a part of the workshops. We have included a service learning component for the workshop leaders.
- I have incorporated it into how I teach the small summer classes.

- We use the ACS course-specific exams to assess our students' performance compared to national norms. Thus, we need to see improvement in the ACS scores to be confident that it is improving performance.
- We just started a similar model in physics this past fall and will be applying it to human anatomy and physiology in fall 2005. PLTL is definitely regarded by our administration as a retention tool.
- It is integrated to work with Calibrated Peer Review (on-line writing assignments) and ChemConnections Modules (chemistry in context).
- We have presented workshops to other schools and departments to help spread interest in PLTL. The reception has been very positive. Our Health Sciences Division has adopted PLTL in a number of their sections after they attended our introductory workshop. We are planning to spread PLTL into other of our Chemistry sections.
- I used PLTL to help train and guide the recitation leaders (RL) hired for me by the Chemistry Department. We use relatively large groups (up to 25) for each RL. The RL divides this larger group into smaller groups of six to eight students. The RL moves from group to group giving encouragement and, when necessary, a nudge in the right direction. Problems to be addressed in the groups are published ahead of time on the course web site. Answers are provided.
- A colleague has implemented PLTL in English and obtained overwhelmingly positive results in terms of student success and retention.

The list reveals a wide range of adaptation, including in-class and lab-related use, coordination with other initiatives, training of recitation leaders, and introduction into other disciplines. Responses of this type made it possible for the evaluator to bring fresh ideas from one site to another, or to make them generally available through publications.

An on-line survey should be thoroughly edited by several people and piloted to be sure that it is easy to use and that responses will provide the kinds of data necessary for analysis.

One of the more important findings from the PLTL experiences is that there is very wide variation in the way college professors approach their teaching responsibilities. The variables include:

- Research commitment in time and energy
- Institutional priorities
- Interest in pedagogy and teaching/learning initiatives
- Scope of teaching assignments and responsibilities for foundation courses
- Career aspirations

These and other factors, such as the type of institution, location, and departmental requirements must be considered by those disseminating an initiative and by evaluators. One apparent universal characteristic among college professors is the desire for as much self-determination as possible. The levels of independence and freedom vary but most professors and instructors believe that the profession promises a reasonably high level of self-determination. A pedagogical initiative that has the potential to increase satisfaction, without reducing self-determination will be much more likely to succeed than one that does not hold such promise.

10.7 Administrative Response

Deans, provosts, and college presidents view educational reform through their own lenses. They are a step removed from the classroom but they also want an institution that is attractive to student applicants, a source of continuing pride to alumni, and fiscally responsible while maintaining high academic standards. Satisfied and successful students are important in attaining these objectives, and consequently administrators encourage and support improvements in teaching and learning.

In order to collect data from administrators about Peer-Led Team Learning, phone or on-site interviews were selected as the most appropriate method. In these interviews administrators were asked about: their views of PLTL in the context of the institution's overall direction; the institution's academic goals; availability of ongoing support; and other local concerns. Questions were clear but intentionally open-ended, so that those interviewed could speak from a broad perspective. The findings of these phone interviews are contained in Chapter 5 on Institutionalization.

The level of administrative involvement in a curricular or pedagogical initiative varies depending on the degree to which the project extends beyond the department. Clearly, a change in textbooks does not require administrative oversight. Changes in a course structure may require departmental approval. Introducing a new course requires a rationale and approval at several levels. PLTL needed various levels of support, depending on the institution and level of implementation. With a WPA grant to cover initial expenses and a small number of students, a department could implement PLTL with little administrative involvement. A dean was often involved when implementation took place on a larger scale with significant organizational needs. Administrators played important roles when PLTL was scaled up or when decisions were made about retaining the program after a period of initial funding.

As a general principle, we encourage evaluators and others gathering data about projects to make an organized effort to involve administrators in appropriate ways—in implementation, adaptation, and evaluation. Their support is critical for sustainability and their influence can be a significant factor in disseminating a project within and beyond an institution.

10.8 Former Leaders Response

The fact that PLTL was studied for over a decade provided an opportunity for the study of the workshop experiences of former peer leaders after they had graduated and had been in their professions or graduate school. A pilot study of the former leaders, at a small school yielded encouraging results and was expanded to a larger group. An online survey was selected as the best instrument for this activity because of the geographic dispersion of the respondents. Chapter 6 on former leaders demonstrates the effectiveness of this study. The analysis was of a type not usually possible with regard to a teaching/learning initiative in that it made connections between particular college experiences, skills acquired, and post-college experiences.

An important lesson from the former leader study is that updated records of alumni are important. Second, good relationships are important. Institutions seemed to get the best responses if they had nurtured and supported students. Finally if a survey is short and easy to answer, more responses are likely.

10.9 Dissemination

The discussion in Chapter 3 on dissemination and diffusion indicates that this is a complex process with a distinct history and literature. By way of lessons and practices applicable to other projects we can use the format developed by the PLTL project. The dissemination of PLTL took place through a four-stage process: (1) providing initial familiarity; (2) providing implementation skills and knowledge; (3) facilitating implementation with project support; (4) encouraging development of new leadership.

The PLTL project used a wide range of strategies to introduce the method. The initial implementers were soon followed by a second and third wave of adopters, many of whom, after experiencing success, became enthusiastic supporters and promoters. The range of avenues used to present and encourage piloting the method, included:

- Informal encounters with colleagues
- Presentations at local, regional, and national conferences
- Meetings sponsored by the PLTL project and grants
- Articles about the program
- The PLTL website

The more complete knowledge and skills required to implement PLTL were generally achieved through two- and three-day workshops sponsored by the project itself, the NSF-supported Chautauqua Faculty Development Program, and by regional networks or individual institutions. These workshops took new implementers through the critical components with emphasis on practical considerations regarding materials, training leaders, scheduling, and other "nuts and bolts," aspects of PLTL.

Implementation of PLTL varied depending on conditions, including the size and scope of the pilot program, local support, expectations of faculty and students, the disciplines involved, and administrative support. From the perspective of dissemination, a key element in implementation at new sites was the availability and effectiveness of support from PLTL practitioners. This assistance included: use of materials developed and published for use with PLTL workshops, training and supervision of peer leaders across cooperating institutions, phone or on-site consultation and trouble shooting, and guidance through the Workshop Project Associate mini-grants.

Emergence of new levels of project leadership took place as adopters took responsibility for PLTL-related projects including: participating in PLTL informational conferences and seminars; using workshops in conjunction with other initiatives such as problem based learning; conducting and publishing studies on the effectiveness of workshops; promoting PLTL at their own and other institutions; developing new approaches to the training of leaders; writing new workshop materials; participating in the PLTL national network.

The evaluation of dissemination is examined and discussed in Chapters 3 and 4. There are a limited number of dissemination and diffusion studies in education. We believe that the analysis presented here, using the work of Rogers, Dearing and others, can be helpful but we strongly encourage more research in this area. In particular, the attributes of innovation—relative advantage, compatibility, complexity, and trialability—provide a framework for considering the likelihood of an initiative being successfully adopted and for monitoring progress after adoption. In addition, college professors are most influenced by opinion leaders who are respected for their scholarship but also viewed as colleagues (Foertsch et al., 1977).

In analyzing the dissemination and adoption of PLTL, we found that a highly significant factor related to the ways in which professors viewed their roles. Sharing responsibilities for teaching and learning with undergraduates was new and challenging to faculty members. As noted at several places in this study, their view of this role change determined a great deal about the likelihood that they would adopt and succeed with PLTL workshops.

10.10 Institutionalization

The term "institutionalization," was frequently used from the start by the PLTL founding group but there was minimal understanding about what would be needed to accomplish it or in fact what it really meant. Rogers' statement that institutionalization is accomplished when an initiative continues beyond the period of initial funding or piloting was an appropriate starting point. An experimental college course may gradually pass through stages until it is accepted within and outside the department as a regular offering. Programs that fund the development of new materials, equipment, or methods of teaching may be said to be institutionalized, after the funding ends, if the materials or processes continue to be used. As has been mentioned several times, reform calculus was institutionalized at many colleges in so far as major elements remained and textbooks that incorporated the approach were adopted.

But Peer-Led Team Learning required an on-going effort, particularly in the selection, training, and supervision of leaders; and changes in faculty generally brought major challenges for the continuation of workshops. Toward the end of the first five-year NSF Workshop Chemistry grant, Gafney and Kampmeier, principal investigator and initiator of PLTL at the University of Rochester, developed a survey to gather preliminary data about how administrators viewed institutionalization.

This survey was followed by an online faculty survey, reported in Chapter 5. Using this survey we were able to make connections among the PLTL implementation, student performance, faculty perceptions about program success, and the likelihood that the program would endure.

Institutionalization is an area that is probably not given enough attention in evaluations and program reviews. Practitioners and evaluators may take one of several approaches: (1) not giving much time or thought to the area because implementation and dissemination take all of their energies; (2) assuming that they are in the process of institutionalizing because the project continues for a few years without incident; (3) believing that institutionalization is essentially out of their hands and so it is not worth worrying about. Each of these ways of thinking has some validity. But in our experience if proponents of a new approach are firm believers in its effectiveness they will find the time and energy to spread the word among their colleagues.

Based on the surveys described above, the results of the WPA mini-grant reports, interviews, and visits, and on a survey of administrators, and analysis of the PLTL project we identified five key elements required for institutionalization: (1) fidelity to the model; (2) funding and administrative support; (3) perceived success; (4) fit with the institution's mission and practice, and (5) a core group of committed faculty.

We conclude that these elements are required, but not necessarily sufficient for a program to be sustained at an institution. We invite others to test these elements, explore the dimensions of each and contribute to the discussion of practices that lead to institutionalization. There are no guarantees that educational initiatives, even those with proven success, will last. We know, however, that sustained success requires cooperation among all participants and a shared commitment to student-centered teaching and learning.

Appendix A
National Science Foundation Support

Peer-Led Team Learning, first under the name Workshop Chemistry, was implemented with the support of the following grants from the National Science Foundation.

1995—NSF Systemic Change Initiative—"A Workshop Chemistry Curriculum"
NSF-DUE 9455920 (1995–2000)

Adapt/Adopt grants
NSF-DUE 9752892, $160,000/Four institutions
NSF-DUE 9950575, $272,162/Seven institutions

National Dissemination Grants
NSF-DUE 99722457, NSF-DUE 0004159 (Supplement) "Peer-Led Team Learning: National Dissemination by the Workshop Project" (1999–2003)
NSF-DUE 0231349, NSF-DUE -0337292 (Supplement)
"PLTL National Dissemination: Building a National Network" (2003–2006)

Multi-Initiative Dissemination
NSF-DUE 0196527 (2000–2004), Super consortium of four "Systemic Change" initiatives

Appendix B
List of PLTL Mini-Grant Recipients: Workshop Project Associates

PLTL workshop project associate grants by institution

Allegheny College
Alma College
Boston University (2)
Brigham Young University
Bronx Community College
Brookdale Community College (2)
Capital University
Central College
Central Michigan University
Coastal Carolina University
College of St.Benedicts/St.John's University
Community College of Rhode Island (2)
Diablo Valley College
Drew University
Eastern Kentuky University
Eastern Oregon University (2)
Emory University
Evergreen Valley College
Florida Atlantic University
Glendale Community College
Howard University
Indian River Community College
Kingsborough Community College
Lane Community College
Le Moyne College
Linfield College
Louisiana Tech University

Mercer University
Miami-Dade Community College (2)
Middlesex Community College
Monroe Community College (4)
Morehouse College
Mount Wachusset Community College
Muskingum College
National Hispanic University
New York City Technical College
Northeastern Illinois University (2)
Northern Michigan University
Ohio University
Pace University
Penn State Schuylkill, The Capital College
Portland Community College (2)
Portland State University
Prince George's Community College (2)
Queensborough Community College
San Jose City College (2)
Sierra College (3)
Southern Illinois University-Edwardsville
Southern Utah University
St. Mary's University
SUNY Brockport
The University of Kansas Center for Research Inc.
University of Miami
University of Alaska Southeast
University of Maine (3)
University of Missouri-St.Louis
University of New Hampshire
University of Oklahoma (2)
University of Oregon
University of Portland (3)
University of Puerto Rico at Cayey
University of Rochester Medical center
Virginia Military Institute (2)
Washington University
Western Oregon University (2)
Whittier College

PLTL is used in many other institutions across the United States and in other parts of the world.

Bibliography

Adelman, C. (1999). *Answers in the tool box: Academic intensity, attendance patterns, and bachelor's degree attainment*. Washington, DC: U.S. Department of Education.

Alger, T. D. (2004). An experiment in improving scores on ACS course-specific examinations at Southern Utah University. *Progressions the PLTL Project Newsletter, 5*(2), 7–10.

Angelo, T. A., & Cross, K. P. (1993). *Classroom assessment techniques: A handbook for college teachers*. San Francisco, CA: Jossey-Bass.

Astin, A. W. (1984). Student involvement: A developmental theory for higher education. *Journal of College Student Personnel, 25*, 297–308.

Astin, A. W. (1993). *What matters in college? Four critical years revisited*. San Francisco, CA: Jossey-Bass.

Astin, A. W., & Astin, H. S. (1993). *Undergraduate science education: The impact of different college environments on the educational pipeline in the sciences* (ERIC Report, NSF, SPA-8955365). Los Angeles, CA: Higher Educational Research Institute. UCLA.

Báez-Galib, R., Colón-Cruz, H., Resto, W., & Rubin, M. R. (2005). Chem-2-chem: A one-to-one supportive learning environment for chemistry. *Journal of Chemical Education, 82*, 1859–1863.

Bean, J. C. (1996). *Engaging ideas: The professor's guide to integrating writing, critical thinking, and active learning in the classroom*. San Francisco, CA: Jossey-Bass.

Berke, T. (2003). Peer-led team learning: An active learning strategy that works. *NEA Higher Education Advocate*. Retrieved December 8, 2006, from http://www2.nea.org/he/advo03/advo1003/front.html

Berliner, D. C. (1988). *The development of expertise in pedagogy (Charles W. Hunt memorial lecture)*. Washington, DC: American Association of Colleges for Teacher Education.

Birmingham, C. & McCord, M. (2004). Group process research: Implications for using learning groups. In Michaelsen, L. K., Knight, A. B. & Fink, L. D. (Eds.), *Team-based learning* (pp. 73–93). Sterling, VA: Stylus.

Black, A. E., & Deci, E. L. (2000). The effects of instructors' autonomy support and students' autonomous motivation on learning organic chemistry: A self-determination theory. *Science Education, 84*, 740–756.

Bodner, G. (1999). Action research: Overcoming the sports mentality approach to assessment/evaluation [Electronic version]. *University Chemistry Education, 3*(1), 31–35.

Bok, D. (1990). *Universities and the future of America*. Durham, NC: Duke University Press.

Bonner II, F. A., & Bailey, K. W. (2006). Enhancing the academic climate for African American college men. In Cuyjet, M. J. & Associates (Eds.), *African American men in college* (pp. 24–46). San Francisco, CA: Jossey-Bass.

Bonsangue, M., & Drew, D. E. (1992). *Long-term effectiveness of the calculus workshop model*. Report to the National Science Foundation. Washington, DC: NSF.

Born, W. K., Revelle, W., & Pinto, L. H. (2002). Improving biology performance with workshop groups. *Journal of Science Education and Technology, 11*, 347–365.

Bottoms, G., Hornig-Fox, J., & New, T. (2001). Research brief: The 2000 high schools that work assessment: Improving urban high schools. Atlanta, GA: Southern Regional Education Board. UR#52CA.

Boud, D., Cohen, R., & Sampson, J. (Eds.) (2001). *Peer learning in higher education: Learning from & with each other*. London: Kogan Page.

Bowen, H. R. (1996). *Investment in learning: The individual and social value of American higher education*. New Brunswick, NJ: Transaction Publishers.

Boyer, E. (1990). *Scholarship reconsidered: Priorities of the professoriate*. Princeton, NJ: Carnegie Foundation for the Advancement of Teaching.

Bradley, A. Z., Ulrich, S. M., Jones, M., & Jones, S. M. (2002). Teaching the sophomore organic course without a lecture. Are you crazy? *Journal of Chemical Education, 79*, 514–519.

Brainard, J. (2007, April 3). America's science test: The tough road to better science teaching. *The Chronicle of Higher Education, 53*(48), 16.

Branson, R. K. (2002). When all the quick fixes fail again, try R&D. Paper presented at the annual meeting of the International Conference on Technology in Education. Retrieved December 11, 2006 from http://www.icte.org/T01_Library/T01_257.PDF

Bruffee, K. A. (1990). *Collaborative learning: Higher education, interdependence, and the authority of knowledge*. Baltimore and London: The Johns Hopkins University Press.

Burke, K. A., & Greenbowe, T. J. (2004). The multi-initiative dissemination project workshops: Who attends and how effective are they? *Journal of Chemical Education, 81*, 897–902.

Chen, H. T. (2005). *Practical program evaluation*. Thousand Oaks, CA: Sage.

Coleman, E. B. (1998). Using explanatory knowledge during collaborative problem solving in science. *Journal of the Learning Sciences, 7*, 387–427.

Coleman, E. B., Brown, A. L., & Rivkin, I. (1997). The effect of instructional explanations on learning from scientific texts. *Journal of the Learning Sciences, 6*, 347–365.

Connolly, M. R., & Millar, S. B. (2006) Using workshops to improve instruction in STEM courses. *Metropolitan Universities Journal, 14*(4), 53–65.

Cooper, J. L., Robinson, P., & McKinney, M. (1994). Cooperative learning in the classroom. In Halpern, D. F. & Associates (Eds.), *Changing college classrooms* (pp. 74–92). San Francisco, CA: Jossey-Bass.

Coppola, B. P. (2007). The most beautiful theories, 2006 James Flack Norris award. *Journal of Chemical Education, 84*, 1902–1912.

Coppola, B. P., Daniels, D. S., & Pontrello, J. K. (2001). Using structured study groups to create chemistry honors sections. In Miller, J., Groccia, J. E., & DiBiasio, D. (Eds.), *Student assisted teaching and learning* (pp. 116–122). Bolton, MA: Anker.

Cracolice, M. (2000). Vygotsky's zone of proximal development. *Progression the PLTL Project Newsletter, 1*(2), 3.

Cracolice, M. S., & Abraham, M. R. (1996). A comparison of computer-assisted instruction, semi-programmed instruction, and teaching assistant-led instruction in general chemistry. *School Science and Mathematics, 4*, 215–221.

Cracolice, M. S., & Deming, J. (2001). Peer-led team learning. *Science Teacher, 68* (1), 20–25.

Cracolice, M. S., & Deming, J. C. (2005). Peer-led team learning: Promoting conceptual understanding and reasoning ability. Paper presented at Winter 2005 CONFCHEM: Trends and new ideas in chemical education. Retrieved December 8, 2006 from http://www.ched-ccce.org/confchem/2005/a/index.html

Cracolice, M. S., & Roth, S. M. (1996). Keller's "old" personalized system of instruction: A "new" solution for today's college chemistry students. *The Chemical Educator 1*(1). Retrieved July 10, 2006 from http://chemeducator.org/newsite.htm

Cress, C. M., Astin, H. S., Zimmerman-Oster, K., & Bukhardt, J. C. (2001). Developmental outcomes of college students' involvement in leadership activities. *Journal of College Student Development, 42*, 15–28.

Creswell, J. (2003). *Research design: Qualitative, quantitative, and mixed methods Approaches*. Thousand Oaks, CA: Sage.

Dearing, J. W. (2004). Improving the state of health programming by using diffusion theory. *Journal of Health Communication, 9*, 21–36.

Deci, W., & Ryan, R. (1990). A motivational approach to the self: Integration in personality. In Dienstbier, R. (Ed.), *Nebraska symposium on motivation: Perspectives on motivation* (Vol. 38, p. 88). Lincoln, NE: University of Nebraska Press.

Dinan, F. J. (2004). An alternative to lecturing in the sciences. In Michaelsen, L. K., Knight, A. B., & Fink, L. D. (Eds.), *Team-based learning* (pp. 97–104). Sterling, VA: Stylus.

Donald, G. D. (2002). *Learning to think.* San Francisco, CA: Jossey-Bass.

Dreyfuss, A. E., & Gosser, D. (2006). In their own words: Learning to be a peer leader. *Student Standpoints About Access Programs in Higher Education.* Retrieved December 8, 2006, from http://education.umn.edu/CRDEUL/monographs.html

Dubinsky, E., Mathews, D., & Reynolds V. E. (1997). Readings in cooperative learning for undergraduate mathematics. In *MAA notes number 44.* Washington, DC: The Mathematical Association of America.

Duch, B. J., Groh, S. E., & Allen, D. E. (Eds.) (2001). *The power of problem-based learningin the twenty-first century.* Sterling, VA: Stylus.

Eberlein, T., Kampmeier, J., Minderhout, V., Moog, R., Platt, T., Varma-Nelson, P., White, H. (2007). Pedagogies of engagement in science: A comparison of PBL, POGIL, and PLTL. *Biochemistry and Molecular Biology Education* (Accepted).

Ege, S. N., Coppola, B. P., & Lawton, R. G. (1997) The University of Michigan undergraduate chemistry curriculum: 1. Philosophy, curriculum, and the nature of change. *Journal of Chemical Education, 74*, 74–83.

Elmore, R. F. (1996). Getting to scale with good educational practice. *Harvard Educational Review, 66*, 1–26.

Erickson, B. L., & Strommer, D. W. (1991). *Teaching college freshmen.* San Francisco, CA: Jossey-Bass.

Etzkowitz, H., Kemelgor, C., & Ussi, B. (2000). *Athena unbound: The advancement of women in science and technology.* New York: Cambridge University Press.

Felder, R. M., & Brent, R. (1994). Cooperative learning in technical courses: Procedures, pitfalls, and payoffs. ERIC document ED 377038. Retrieved April 18, 2007 from http://www.ncsu.edu/felder-public/Papers/Coopreport.html

Fink, L. D. (2003). *Creating significant learning experiences.* San Francisco, CA: Jossey-Bass.

Finkel, L., & Monk, G. S. (1997). Teachers and learning groups: Dissolution of the Atlas complex. In Dubinsky, E., Mathews, D., & Reynolds, V. E. (Eds.), *Readings in cooperative learning for undergraduate mathematics* (pp. 5–12).Washington, DC: The Mathematical Association of America.

Foertsch, J., Millar, S. B., Squire, L., & Gunter, R. (1997). Persuading professors: A study of the dissemination of educational reform in research universities. The LEAD Center, University of Wisconsin-Madison. Retrieved February 12, 2007 from http://homepages.cae.wisc.edu/~lead/pages/products/iadrpt.pdf

Ford, R. (2005). High school profiles of mathematically underprepared college freshmen. Online preliminary report, California State University, Chico. Retrieved January 18, 2007 from http://www.csuchico.edu/eapapp/PDFfiles/CollegeReadiness.htm

Gafney, L. (2000a). Evaluation strategies. *Progressions the PLTL Project Newsletter, 1*(2), 5–7.

Gafney, L. (2000b). Evaluation of dissemination efforts: Faculty ready to adopt PLTL. *Progressions the PLTL Project Newsletter, 1*(4), 1,17–18.

Gafney, L. (2001a). Evaluating student performance. *Progressions the PLTL Project Newsletter, 2*(2), 3–4.

Gafney, L. (2001b). The impact of research on undergraduates' understanding of science. *The Council on Undergraduate Research Quarterly, 21*, 172–176.

Gafney, L. (2001c). Workshop evaluation. In Gosser, D. K., Cracolice, M. S., Kampmeier, J. A., Roth, V. Strozak, V. S., & Varma-Nelson, P. (Eds.), *Peer-led team learning: A guidebook* (pp. 75–92). Upper Saddle River, NJ: Prentice Hall.

Gafney, L. (2005). Student and faculty views of the research mentor/teacher. *Journal of College Science Teaching, 34*(6), 52–56.

Gafney, L., & Varma-Nelson, P. (2002). What happens next? A follow-up study of workshop leaders at St. Xavier University. *Progressions the PLTL Project Newsletter, 3*(2), 1, 8–9.

Gafney, L., & Varma-Nelson, P. (2007). Evaluating peer-led team learning: A study of long-term effects on former workshop peer leaders. *Journal of Chemical Education, 84*, 535–539.

Gaines, M. (2001). A case study: University of Miami. *Progressions the PLTL Project Newsletter,* 2(2), 14.

Geiger, R. L. (1994). *Research and relevant knowledge: American research universities since World War II*. New York: Oxford University Press.

Goroff, N. (1998). Report on: Workshop chemistry. Retrieved March 20, 2006, from *Chemical Educator Online 3(1)*: http://chemeducator.org/bibs/0003001/00030172.htm

Goroff, J. P., Bertenthal, M. W., Labov, J. B., & Curtis, P. C. (Eds.) (2002). *Learning and understanding: Improving advanced study of mathematics and science in U. S. high schools*. Washington, DC: National Academy Press.

Gosser, D. K. (2008). PLTL in general chemistry: Scientific learning and discovery. In Pienta, N., Cooper, M., & Greenbowe, T. (Eds.), *The Chemists' guide to effective teaching* (Vol. 2). Upper Saddle River, NJ: Prentice Hall (in press).

Gosser, D. K., & Roth, V. (1998). The workshop chemistry project: Peer-led team learning. *Journal of Chemical Education, 75*, 185–188.

Gosser, D., K., Cracolice, M. S., Kampmeier, J. A., Roth, V. Strozak, V.S., & Varma-Nelson, P. (2001). *Peer-led team learning: A guidebook* (pp. 75–92). Upper Saddle River, NJ: Prentice Hall.

Gosser, D. K., Strozak, V., & Cracolice, M. (2006). *Peer-led team learning: General Chemistry* (2nd ed.). Upper Saddle River, NJ: Prentice Hall.

Griswold, J., Gaines, M., Baronas-Lowell, D., Van Vliet, K., Mallery, C., Stetler, D. Gaydos, B., & Wilson, D. (2005). Introductory biology: Modules 1 to 12. *Progressions the PLTL Project Newsletter, 7*(1).

Gross, N., Giacquinta, J. B., & Bernstein, M. (1971). *Implementing organizational innovations*. New York: Basic Books.

Halpern, D. F. (1994). Rethinking college instruction for a changing world. In Halpern, D. F. & Associates (Eds.), *Changing college classrooms* (pp. 1–12). San Francisco: Jossey-Bass.

Hanson, D. (2006). *Instructor's guide to process-oriented guided inquiry*. Lisle, IL: Pacific Crest.

Hanson, D., & Wolfskill, T. (2000). A new model for instruction. *Journal of Chemical Education, 77*, 120–130.

Hewlitt, J. A. (2004). In search of synergy: Combining peer-led team learning with the case study method. *Journal of College Science Teaching, 33*(4), 28–31.

Houston, J. S. (2007, May 31). Future skills demands, from a corporate consultant perspective. Presented at a workshop on Research Evidence Related to Future Skill Demands, at the National Academies Center.

Hunter, A. -B., Laursen, S. L., & Seymour, E. (2006). Becoming a scientist: The role of undergraduate research in students' cognitive, personal and professional development. *Science Education, 91*(1), 36–74.

Hutchings, P., & Shulman, P. (1999). The scholarship of teaching: New elaborations, new developments. *Change, 31*(5), 10–15.

Jaques, D. (2000). *Learning in groups* (3rd ed.). Sterling, VA: Stylus.

Johnson, D. W., Johnson, R. T., & Smith, K. A. (1991). *Active learning: Cooperation in the college classroom*. Edina, MN: Interaction Book.

Kampmeier, J. A., & Varma-Nelson, P. (2008). Peer-led team learning; Organic. InPienta, N., Cooper, M. M., & Greenbowe, T. (Eds.), *Chemists' guide to effective teaching* (Vol. 2).Upper Saddle River, NJ: Prentice Hall (in press).

Kampmeier, J. A., Wamser, C. C., Varma-Nelson, P., & Wedegaertner, D. (2006). *Peer-led team learning: Organic chemistry* (2nd ed.). Upper Saddle River, NJ: Prentice Hall.

Katzenback, J. R., & Smith, D. K. (1999). *The wisdom of teams*. New York: Harper Collins.

Kimbrough, W. M., & Harper, S. R. (2006). African American men at historically black college and universities: Different Environments, similar challenges. In Cuyjet, M. J. (Ed.), *African American men in college* (pp. 189–209). San Francisco, CA: Jossey-Bass.

King, A. (1993). From sage of the stage to guide on the side. *College Teaching, (41)*. Retrieved February 7, 2007 fromhttp://www.questia.com/PM.qst?a=0&se=gglsc&d=94305197

Kuh, G. D. (2003). What we're learning about student engagement from NSSE. *Change*, 35(2), 2433.

Kuh, G. D. (2006). 'Engagement' and the underprepared. *Inside Higher Ed*. Retrieved January 18, 2007 from http://insidehighered.com/news/2006/08/01/engage

Landis, R. B. (2005). Retention by design: Achieving excellence in minority engineering education. Retrieved July 12, 2006 from: http://www.google.com/search?hl=en&lr=lang_en&ie=ISO-8859-1&q=Retention + by + Design&btnG=Search

Larson, R. S., & Dearing J. W. (2002). Diffusion of best practices by grantors. Retrieved July 12, 2006 from: http://www.google.com/search?hl=en&lr=lang_en&ie=ISO-8859-1&q=Diffusion + of + Best + Practices + by + Grantors&btnG=Search

Larson, R. S., & Meyer, G. (2007). Diffusing STEM pedagogies: The role of opinion leaders. *Metropolitan Universities Journal, 17(4)*, 77–91.

Lewis, J. (2004). Peer-led guided inquiry: Combining systemic change models. *The PLTL Project Newsletter, 5*(2), 3–4.

Lewis, S. E., & Lewis, J. E. (2005). Departing from lectures: An evaluation of a peer-led guided inquiry alternative. *Journal of Chemical Education, 82*, 135–139.

Light, R. J. (2001). *Making the most of college: Students speak their minds*. Cambridge: Harvard University Press.

Lopatto, D. (2004). What undergraduate research can tell us on research on learning. *Project Kaleidoscope Volume IV*. Retrieved January 19, 2007 from http://www.pkal.org/documents/Vol4WhatUndergradResearchCanTellUs.cfm

Lovitts, B. E. (2001). *Leaving the ivory tower: The causes and consequences of departure from doctoral study*. Lanham, MD: Rowman & Litlefield.

Lyle, K. S., & Robinson, W. R. (2003). A statistical evaluation: Peer-led team learning in an organic chemistry course. *Journal of Chemical Education, 80*, 132–134.

Maxfield, M. (2001). The view from industry. In Gosser, D. K., Cracolice, M. S., Kampmeier, J. A., Roth, V., Strozak, V. S., & Varma-Nelson, P. (Eds.), *Peer-led team learning: A guidebook* (pp. 75–92). Upper Saddle River, NJ: Prentice Hall.

Mazur, E. (1997). *Peer instruction: A user's manual*. Upper Saddle River, NJ: Prentice Hall.

McCreary, C. L., Golde, M. F., & Koeske, R. (2006). Peer Instruction in the general chemistry laboratory: Assessment of student learning. *Journal of Chemical Education, 83*, 804–810.

McKeachie, W. J., & Svinicki, M. (2006). *McKeachie's teaching tips*. Boston, MA: Houghton Mifflin.

Michaelsen, L. (2004). Getting started with team-based learning. In Michaelsen, L. K., Knight, A. B. & Fink, L. D. (Eds.), *Team-based learning* (pp. 27–50). Sterling, VA: Stylus.

Michaelsen, L. K. & Knight, A. B. (2004). Creating effective assignments. In Michaelsen, L. K., Knight, A. B. & Fink, L. D. (Eds.), *Team-based learning* (pp. 51–72). Sterling, VA: Stylus.

Michaelsen, L. K., Knight, A. B., & Fink, L. D. (Eds.) (2004). *Team-based learning*. Sterling, VA: Stylus.

Miller, J. E., Groccia, J. E., & Miller, M. S. (Eds.) (2001). *Student-assisted teaching: A guide to faculty student teamwork*. Bolton, MA: Anker.

Miller, L. S. (1995). *An American imperative: Accelerating minority educational advancement*. New Haven, CT: Yale University Press.

Millis, B. J., & Cottell, P. G. (1998). Cooperative learning for higher education faculty. Phoenix, AZ: Oryx.

Moog, R. (2004). Process-oriented guided inquiry learning, Retrieved April 7, 2007 from:http://www.pogil.org

Moog, R., Creeegan, F. J., Hanson, D. M., Spencer, J. N., & Straumanis, A. R. (2006). Process-oriented guided inquiry learning: PPOGIL and the POGIL project. *Metropolitan Universities Journal, 14*(4), 41–52.

Nakhleh, M. B. Lowrey, K. A., & Mitchell, R. C. (1996). Narrowing the gap between concepts and algorithms in freshman chemistry. *Journal of Chemical Education, 73*, 758–762.

National Science Foundation web site, Retrieved April 20, 2007 from http://www.nsf.gov/statistics/

Newman, F., & Holzman, L. (1993). *Lev Vygotsky: Revolutionary scientist*. New York: Routledge.

Perry, W. (1970). *Forms of intellectual and ethical development in the college years*. New York: Holt, Rinehart, & Winston.

Pink, D. H. (2005). *A whole new kind of mind*. New York: Berkley.

Platt, T., Barber, E., Yoshinaka, A., & Roth, V. (2003). Problem-based learning: An innovative selection and training program for problem-based learning (PBL) workshop leaders. *Biochemistry and Molecular Biology Education, 31*, 132–136.

Progressions the PLTL Project Newsletter. Available online at http://www.pltl.org

Rait, E. (1995). Against the current: Organizational learning in schools. In Bacharach, S. B., & Mundell, B. (Eds.), *Images of schools: Structures and roles in organizational behavior* (p. 98). Thousand Oaks, CA: Corwin.

Rogers, E. M. (2003). *Diffusion of innovations* (5th ed.). New York: Free Press.

Roth, V., Goldstein, E., & Marcus, G. (2001). *Peer-led team learning: A handbook for leaders*. Englewood Cliffs, NJ: Prentice Hall.

Sarquis, J. L., & Detchon, J. C. (2005). The PLTL experience at Miami University. *Project Kaleidoscope Volume IV*. Retrieved April 5, 2007 from:
http://www.pkal.org/documents/Vol4PeerLedTeamLearning.cfm

Sarquis, J. L., Dixon, L. J., Gosser, D. K., Kampmeier, J. A., Roth, V., Strozak, V. S., & Varma-Nelson, P. (2001). The workshop project: Peer-led team learning in chemistry. In Miller, J. E., Groccia, J. E., & Miller, M. S. (Eds.), *Student-assisted teaching: A guide to faculty student teamwork* (pp. 150–155). Bolton, MA: Anker.

Schommer, M. (1990). Effects of beliefs about the nature of knowledge on comprehension. *Journal of Educational Psychology, 82*, 498–504.

Seymour, E. (2005). Student assessment of learning gains. Retrieved September 10, 2005 from: http://www.wcer.wisc.edu/salgains/instructor/

Seymour, E., & Hewitt, N. M. (1997). *Talking about leaving: Why undergraduates leave the sciences*. Boulder, CO: Westview.

Seymour, E., Hunter, A. B., Laursen, S. L., & DeAntoni, T. (2004). Establishing the benefits of research experiences for undergraduates: First findings from a three-year study. *Science Education, 88*, 493–534.

Scheirer, M. A. (2005). Is sustainability possible? A review and commentary on empirical studies of program sustainability. *American Journal of Evaluation, 26*, 320–347.

Spencer, J. N. (1999). New directions in teaching chemistry: A philosophical and pedagogical basis. *Journal of Chemical Education, 76*, 566–569.

Steele, C. (1999). Thin-ice: Stereotype threat and black college students. *Atlantic Monthly, 248*(2), 44–54.

Steele, C. (2003). Stereotype threat and African-American student achievement. In Perry, T., Steele, C., & Hilliard III, A. (Eds.), *Young, gifted and black* (pp. 109–130). Boston, MA: Beacon.

Stewart, B. N., Amar, F. G., & Bruce, M. R. M. (2007). Challenges and rewards of offering peer-led team learning in a large general chemistry course. *Australian Journal of Chemistry, 67*, 31–36.

Tenney, A., & Houck, B. (2004). Learning about leadership: Team learning's effect on peer leaders. *Journal of College Science Teaching, 33*, 25–29.

Tien, L. T., Roth, V., & Kampmeier, J. A. (2002). Implementation of a peer-led team learning instructional approach in an undergraduate organic chemistry course. *Journal for Research in Science Teaching, 37*, 606–632.

Tien, L. T., Roth, V., & Kampmeier, J. A. (2004). A course to prepare peer leaders to implement a student-assisted learning method. *Journal of Chemical Education, 81*, 1313–1321.

Tobias, S. (1990). *They're not dumb, they're just different—stalking the second tier*. Tucson, A Z: Research Corporation.

Tobias, S., & Raphael, J. (1997). *The hidden curriculum: Faculty-made tests in science: Part I Lower division courses*. New York: Plenum.

Topping, K. (1998). Tutoring. The International Academy of Education: Educational Practices Series—5. Retrieved April 5, 2007 from http://unesdoc.unesco.org/images/0012/001254/125454e.pdf

Treisman, P. M. (1985). *A Study of the mathematics performance of black students at the University of California, Berkeley*. Unpublished doctoral dissertation, University of California, Berkeley.

Treisman, P. Uri. (1992) Studying students studying calculus: A look at the lives of minority mathematics students in college. *The College Mathematics Journal, 23*(5), 362–372.

Tucker, A. C., & Leitzel, J. R. (1995). *Assessing calculus reform efforts*. Washington, DC: The Mathematical Association of America.

UMKC: Supplemental Instruction (2006). Retrieved January 12, 2006 from http://www.umkc.edu/cad/si/

Varma-Nelson, P. (2004). The peer-led team learning workshop model. *Project Kaleidoscope Volume IV*. Retrieved April 5, 2007 from: http://www.pkal.org/documents/Vol4PeerLedTeamWorkshop.cfm

Varma-Nelson, P. (2006). Peer-led team learning: The model, its dissemination and its future. *Metropolitan Universities Journal, 14*(4), 19–29.

Varma-Nelson, P., & Coppola, B. P. (2004). Team learning. In Pienta, N., Cooper, M. M., & Greenbowe, T. (Eds.), *The chemists' guide to effective teaching*. Upper Saddle River, NJ: Prentice Hall.

Varma-Nelson, P., & Cracolice, M. (2001). *Peer-led team learning: General, organic, and biological chemistry*. Upper Saddle River, NJ: Prentice Hall.

Varma-Nelson, P., & Dreyfuss, A. W. (2003). PLTL dissemination manual. Retrieved January 17, 2007 from http://www.sci.ccny.cuny.edu/~chem wksp/DisseminationManual.html

Varma-Nelson, P., & Gosser, D. K. (2005). Dissemination of peer-led team learning (PLTL) and formation of a national network embracing a common pedagogy. In Ouellet, M. (Ed.), *Teaching inclusively: Diversity and faculty development* (pp. 503–518). Stillwater, OK: New Forums Press.

Varma-Nelson, P., Cracolice, M. S., & Gosser, D. K. (2003). Peer-led team learning: A student-faculty partnership for transforming the learning environment, *Successful Pedagogies*. Retrieved April 5, 2007 from: http://www.aaas.org/publications/books_reports/CCLI/

Vegso, J. (2005). CRA Taulbee trends: Female students & faculty. Computing Research Association. Retrieved December 9, 2006 from http://www.cra.org/info/taulbee/women.html

Wamser, C. C. (2006). Peer-led team learning (PLTL) in organic chemistry: Student performance, success, and persistence in the course. *Journal of Chemical Education, 83*, 1562–1566.

Weimer, M. (2002). *Learner-centered teaching*. San Francisco, CA: Jossey-Bass.

Whitehead, B. D. (2006, May 6). The boy problem. *Commonweal, 133*, 7.

Wholey, J. S. (1994). Assessing the feasibility and likely usefulness of evaluation. In Wholey, J. S., Hatry, H. P., & Newcomer, K. E. (Eds.), *Handbook of practical program evaluation* (pp. 15–39). San Francisco, CA: Jossey-Bass.

Woodward, A., Gosser, D., & Weiner, M. (1993). Problem-solving workshops in general chemistry. *Journal of Chemical Education, 70*, 651–665.

Index

Bold page-type denotes referenced titles; lower case t(s) following a page number denote a table or tables.